T0290708

# High-Lift Aerodynamics

# High-Lift Aerodynamics

Jochen Wild

CRC Press
Taylor & Francis Group
Boca Raton London New York

CRC Press is an imprint of the
Taylor & Francis Group, an **informa** business

First edition published 2022
by CRC Press
6000 Broken Sound Parkway NW, Suite 300, Boca Raton, FL 33487-2742

and by CRC Press
4 Park Square, Milton Park, Abingdon, Oxon, OX14 4RN

ISBN: 978-1-032-11546-7 (hbk)
ISBN: 978-1-032-11559-7 (pbk)
ISBN: 978-1-003-22045-9 (ebk)

DOI: 10.1201/9781003220459

Typeset in Times
by KnowledgeWorks Global Ltd.

# High-Lift Aerodynamics

*Von der Fakultät für Maschinenbau
der Technischen Universität zu Braunschweig
angenommene Habilitationsschrift
zur Erlangung der Venia legendi für das Lehrgebiet
„Aerodynamik"*

*von Dr.-Ing. Jochen Wild
aus Bad Tölz*

*4. Mai 2021*

*Habilitation thesis accepted by
the Faculty of Mechanical Engineering
of the Technical University Braunschweig
for obtaining the Venia Legendi for the subject area
"Aerodynamics"*

*by Dr.-Ing. Jochen Wild
born in Bad Tölz, Bavaria, Germany*

*May 4th, 2021*

# Dedication

*to Inge*

# Contents

# Preface

*High-Lift Aerodynamics* deals with the aerodynamic behavior of lift augmentation means, mainly named high-lift devices. It is a subset of the aerodynamics of flying vehicles at relatively low speed. Within this work, the focus is put on civil transport aircraft application, although the associated physics apply to any other aircraft, too.

Nevertheless, the naming *High-Lift Aerodynamics* does not fully describe the aim of the associated technology. High-lift devices are mainly applied to reduce the speed of an aircraft during take-off and approach. Since the aerodynamic lift force in steady non-accelerated flight is equal to the weight of the aircraft, it is not directly the lift force that has to be increased but the aerodynamic lift coefficient – and therefore the minimum speed to the maximum achievable lift coefficient. The correct naming of the subject would therefore be *High Maximum Lift Coefficient Aerodynamics*. But this terminology has never been used for simplicity.

Even though the aim is to reduce the flight speed, the terminology *Low-Speed Aerodynamics* – although sometimes used – is misleading. It is associated with real slow flows at velocities of a few meters per second, where friction forces are of much higher importance. The speed regime addressed in *High-Lift Aerodynamics* covers relatively low speeds in relation to the speed regime of cruising aircraft. Typical flow velocities at landing range from 30 kts for very small airplanes up to 200 kts for large transport airplanes.

High lift coefficients are often related to flow conditions at relatively high angles of attack, the angle between the aircraft axis, and the flow direction. Nevertheless, the terminology *High-Angle-of-Attack Aerodynamics* is closely related to the specific flow over slender delta wings and the associated vortex dominated flows. This type of flow is not the one in mind for this book, although vortices itself can be used to augment the lift capabilities of lifting surfaces.

Within this book, the topic of high-lift aerodynamics is approached from different directions. After an introductory chapter, a first view discusses the physical limits of lift generation giving the lift generation potential. Second, it is discussed what is needed to get an aircraft flying safely by analyzing the high-lift-related requirements for certifying an aircraft. And third, the needs of an aircraft are analyzed to improve its performance during take-off, approach, and landing.

After this orientation, the different mechanisms to increase the lift coefficient are discussed. Throughout this book, a differentiation is made between passive and active high-lift. *Passive High-Lift* means that the change in lift is obtained only by modifying shapes and thereby redirecting the flow. In contrast, *Active High-Lift* means to change the energy content of the flow for the same purpose.

To round up the topic of high-lift aerodynamics, the last part of the book describes methods to evaluate and design high-lift systems in an aerodynamic sense. It covers in brief numerical as well as experimental simulation methods. A special chapter is dedicated to the aerodynamic design of high-lift systems. It is obvious, that these topics cannot be covered in full scope here. The aim is therefore to concentrate on the relation to the specifics of high-lift aerodynamics and to provide a first insight.

For more details on simulation and design methods, the honored reader is referred to specialized literature.

As the reader will see later on, the topic of high-lift aerodynamics touches several different fields and disciplines in aircraft technology. As each of these have their own standards and naming conventions, it is difficult to exclude doubled meanings of symbols. E.g., $T$ is used for temperature in fluid mechanics but for thrust in flight mechanics. To make the book in this sense more readable, a general nomenclature is omitted. Instead, a separate nomenclature is given for each specific chapter.

# Author

**Jochen Wild, PhD,** studied aerospace and mechanical engineering at Technische Universität München, Germany, and Technische Hochschule Darmstadt, Germany, from which he earned the Diploma of Mechanical Engineering in 1995. He earned a doctorate Dr.-Ing. at Technische Universität Braunschweig in 2001, Germany, where he has been lecturing since 2014. He joined the German Aerospace Center DLR (Deutsches Zentrum für Luft- und Raumfahrt e.V.) in 1995, first as a PhD student, later on as a research engineer and research scientist at the Institute of Design Aerodynamics, now Institute of Aerodynamics and Flow Technology, where he headed the research group of High-Lift Aerodynamics within the Transport Aircraft Department from 2007 to 2018. He is a specialist in high-lift aerodynamics with emphasis on design and validation. His further research interests include numerical optimization, CFD, mesh generation, wind tunnel testing, and wind energy.

# 1 Introduction

## NOMENCLATURE

| | | | | | |
|---|---|---|---|---|---|
| $A_{ref}$ | $m^2$ | Reference wing area | $R$ | $J/kg\ K$ | Specific gas constant (287.058 for dry air) |
| $a$ | $m/s$ | Speed of sound | $T$ | $K$ | Temperature |
| $D$ | $N$ | Aerodynamic drag force | $V_\infty$ | $m/s\ (kts)$ | Flight speed/flow velocity |
| $H$ | $m\ (ft)$ | Flight altitude | $W$ | $N$ | Weight force |
| $L$ | $N$ | Aerodynamic lift force | $\rho_\infty$ | $kg/m^3$ | Air density |
| $M_\infty$ | – | Flight Mach number | $\gamma$ | $°\ (deg)$ | Flight path angle |
| $n_z$ | – | Load factor | $\kappa$ | – | Isentropic exponent (1.4 for ideal gas) |
| $p_\infty$ | $N/m^2$ | Static pressure | | | |

The major motivation of lift augmentation is the reduction of the flight speed in the critical flight phases in proximity to the ground, namely the take-off, climb, approach, and landing phases. It is the well-known general relation between the weight of the aircraft and the lift generated being approximately equal for steady non-accelerated flight

$$W = L\cos\gamma - D\sin\gamma \quad \Rightarrow \quad W \approx L \text{ for } \gamma < 10° \tag{1.1}$$

that links the speed reduction to the increase of lift coefficient

$$L = C_L \frac{1}{2}\rho_\infty V_\infty^2 A_{ref} = C_L \frac{1}{2}\kappa p_\infty M_\infty^2 A_{ref}. \tag{1.2}$$

According to eq. (1.2), the ways to achieve lower flight speeds are:

a. to reduce the aircraft mass;
b. to increase the maximum lift coefficient;
c. to increase the wing area; and
d. to increase the density of the fluid.

The first option is feasible but directly affects the payload and fuel capacity. On the other hand, for cruise flight, it is known that the efficiency of an aircraft is increased by a higher wing loading. Positively, it can be stated that the minimum speed of a transport aircraft during landing is in general lower than for take-off due to the consumed fuel.

The second option is the most prominent way to vary the minimum flight speed by high-lift systems.

DOI: 10.1201/9781003220459-1

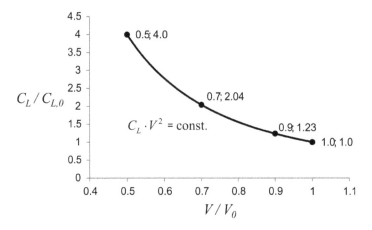

**FIGURE 1.1**   Relation of speed reduction with lift coefficient increase at constant aircraft mass and wing area.

The third option has been tried in several experimental projects but turns out to be mechanically very challenging. Nevertheless, some high-lift systems make use of a wing area increase, e.g., the Fowler flap. In order to not mix the area effect with other aerodynamic effects, the wing reference area is kept constant from the clean wing for high-lift system design. So, any aerodynamically effective change, including area increase by a high-lift system, is attributed to the lift coefficient.

The last option is not known how to be controllable. But it shows that climatic conditions have an influence on the minimum flight speed of an aircraft.

Figure 1.1 shows the dependency of flight speed and lift coefficient for constant lift force. The quadratic relation of lift coefficient and speed (1.2) has consequences for the order of magnitude of increasing the lift coefficient to get a certain speed reduction. A 10% increase in lift coefficient results in less than 5% speed reduction. To reduce the speed by 10%, the lift coefficient has to be increased by 23%. Doubling the lift coefficient will reduce the speed by approx. 30%. To halve the flight speed, it would need four times the lift coefficient. These numbers show that, for a significant reduction of flight speed, a great effort has to be made.

Now, porting this relation to a typical wide-body transport aircraft – e.g., of an Airbus A330 type with close to 200 tons of maximum landing weight and a wing area of about $A_{\text{ref}} \approx 350 \ m^2$ – leads to a situation as depicted in Figure 1.2. Respecting a safety margin towards the real maximum lift coefficient as outlined in Chapter 3 and anticipating the corridor of the approach speed at large airports for this type of aircraft that will be discussed in Chapter 4, it is seen that for achieving the ICAO approach speed corridor, the required maximum lift coefficient is about a factor of 2 to 3 higher than what can be achieved by a plain swept wing.

Skipping the approach speed corridor, it is in principle possible to fly at higher approach speeds. It is only required to find an airport to land on. Using the same wide-body aircraft as before, Figure 1.3 shows the landing distance of such an aircraft

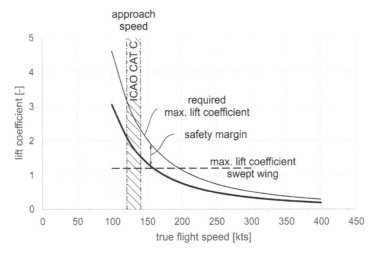

**FIGURE 1.2** Required maximum lift for a wide-body aircraft in relation to the target approach speed.

in relation to the approach speed. It must be noted that according to requirements outlined in Chapter 3, the required airport field length is established in a way so that the aircraft must be able to land within 60% of the available field length. On the right-hand side, a statistic of the available field length of medium and large airports is shown. While staying within the ICAO approach speed corridor, it is possible to land on 90% of such airports. At an approach speed of 200 kts required by the plain wing maximum lift coefficient, the field length of only 30% of the airports is still sufficient. In the left figure, the longest landing field is marked, which is Zhukowski

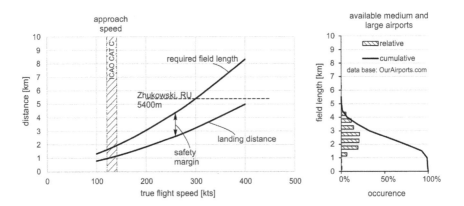

**FIGURE 1.3** Required landing field length of a wide-body aircraft in relation to the approach speed (right) and the occurrence count of medium and large airports by field length (left). (Data from [1].)[1]

Airport near Moscow in Russia with a runway length of 5.4 km. It would be the only airport to land such a wide-body aircraft with an approach speed of 300 kts.

## 1.1   A SHORT HISTORY OF HIGH-LIFT SYSTEMS

Lift augmentation has been the focus of aircraft development since the early years of flight. Already in the view of the early pioneers of flight, take-off and landing were the riskiest phases of flight. The simple relation that the impact of failures scales dramatically with the current speed led to the aim to reduce the flight speed in these phases.

Nevertheless, the foundation for the topic of high-lift aerodynamics has been laid nearly at the same time as the first flights of the Wright brothers. And it was the foundation of active flow control that came first. In 1904, Ludwig Prandtl published his famous work "on fluid motion at very low viscosity" [2] where he laid the foundation of boundary layer theory. In this work, Prandtl made an experiment that proved that his discovered boundary layer is responsible for flow separation. Figure 1.4 shows the original figures of the experimental setup. His experiment comprised a cylinder with a small slot in his famous water tunnel. In the first series of experiments, he analyzed the flow and saw the laminar double bubble at low flow velocities (upper left) and turbulent separation at higher velocities (upper right). Afterwards, he rotated the cylinder with the slot now being at a 45° position and sucked water out of the interior. In his explanation, by this, he removed the boundary layer and showed the flow being attached for laminar and turbulent flows (lower row). From this, he concluded that the existence of the boundary layer is responsible for flow separation. He did not know yet that this was also the foundation for active flow control by boundary layer suction and therefore the first principle of lift augmentation.

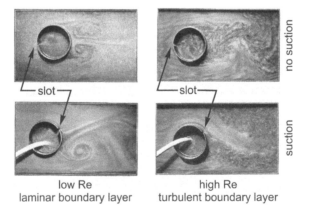

FIGURE 1.4   Prandtl's experiments with a cylinder to show the role of the boundary layer for flow separation. (Adaption of figure 11 in Prandtl [2].)

### 1.1.1 DEVELOPMENT OF HIGH-LIFT DEVICES

After the first flight of Orwell and Wilbur Wright, a significant number of early pioneers of flight tried to build new flying machines – and failed to land them safely. It was a common observation that when getting the aircraft too slow during landing, it stalls, gets uncontrollable, and crashes. It did not take long until a focus was put on reducing the flight speed when landing without getting into the risk of stall. Nayler, Stedman, and Stern were first to propose to hinge the wing to adapt the camber of the airfoil [3]. They experimented between 1912 and 1914 with a hinged wing and called it plain flap. Figure 1.5 shows the used RAF 9 airfoil with the investigated flap slightly larger than 1/3rd of chord. In fact, this was not only the beginning of what we call high-lift devices today. It was the beginning of any today's control surface, whether it is an aileron, elevator, or rudder.

Already at the end of World War I, one of the most dramatic stories in high-lift system development took place. After an accident, the German Gustav Lachmann gave detailed thoughts on reducing the stall speed by increasing the lift coefficient up to a previously unknown value. With detailed theoretical work using potential theory, he came up with the idea to introduce slots into the wing. In 1918, Lachmann filed the patent on slotted wings [4] at the Deutsches Reichspatentamt. But there, it was simply not believed that adding slots to a wing could increase its lift capability – and the patent was rejected.

In parallel and completely independent, Frederic Handley Page experimented with slotted wings, placing small vanes ahead of the main wing with a very significant effect on the stall speed. He filed his patent in 1920 [5]. From there on, what we call a slat today is therefore associated with his name. At the end, the license fees on the slat principle were a much bigger economic success for Handley Page than the production of airplanes.

The happy end: After getting his doctoral degree based on his theoretical work on slotted airfoils in 1923, Lachmann moved to Handley Page's aircraft company and became head of the research department. He stayed there for the rest of his working career. Nevertheless, his name is today more known for his work on boundary layer control while the slat device is still attributed to Handley Page.

Only a few years later, it was the first invention of Harlan D. Fowler to propose a wing area increase to reduce the stall speed. In his patent from 1921, the airfoil was variable in the chordwise direction [6]. It took him further six years to develop what is today called the Fowler flap [7]. In this concept, he combined the wing area increase with the camber increase and a slot. In fact, the Fowler flap has been up to now one of the most important inventions in aircraft development. No large transport aircraft today is flying without a flap based on Fowler's principle. The aircraft we know today would not have been possible without it.

**FIGURE 1.5**    Hinged plain flap investigated by Nayler, Stedman, and Stern. RAF 9 airfoil with a 0.385c plain flap tested in 1912–1913. (Figure 2 in Smith [13].)

It took another 12 years until the next device came up. Ludwig Bölkow looked for a mechanism to open and close the slot for the slat device during flight. He came up with a solution, where the slot was opened by drooping a part of the leading edge downwards [8]. Only in the second instance, he proposed to move the complete leading edge – the birth of the droop nose today flying at Airbus A350 XWB and A380 aircraft.

The last device principle invented was the nose split flap. During World War II, Werner Krüger experimented with a panel deflected at the leading edge from the lower side against the flow direction [9]. This device is today simply known as the Krueger flap. Krueger made something different than his ancestors except Nayler, Stedman, and Stern. Instead of issuing a patent, he published his invention in a public report – a decision with late consequences. With the beginning of the age of jet airliners in the late 1950s, Boeing relied completely on using Krueger flaps at the leading edge. But the first patent on a Krueger device was not filed before 1972 by Frederick T. Watts [10]. After getting his patent verified, he claimed license fees from Boeing. In the course of the legal dispute, finally Boeing hired Werner Krueger for consulting. Werner Krueger was able to prove that, due to his publication, the device has to be seen as common knowledge and a patent on the device itself is worthless.

By 1943, all known types of high-lift devices had been invented. There are two descriptive summaries of known high-lift devices from that age [11, 12] that describe all these systems with its way of working as we know them today – with one exception: The aerodynamics are all addressed to camber increase and the control of the boundary layer, in case of slotted wings by starting new fresh boundary layers. This understanding goes back again to the work of Prandtl from 1904. Weyl in his summary of 1945 [12] succeeds to find a physical explanation for every device that brings everything back to boundary layer control – a misunderstanding that survived partly up to today.

It took 60 years from the first high-lift device until A.M.O. Smith – Chief Aerodynamics Engineer for Research at the Douglas Aircraft Company – gave his famous "Wright Brother's lecture" in 1974, simply named "High-Lift Aerodynamics" [13]. He was the first able to summarize the complete flow physics associated with high-lift flows and especially with slotted wings. He really emphasized that the influence of multiple lifting surfaces on each other has much stronger effects on the lift generation than what could be explained by boundary layer stabilization. Earlier works – as those of Lachmann [4], Betz [14], and Pleines [15] – had the ideas in them but have not been valued enough. Smith was able to summarize this in a clear and complete way and his corresponding journal article is the textbook of high-lift aerodynamics.

### 1.1.2 TRENDS FOR LARGE TRANSPORT AIRCRAFT

Looking at today's flying large transport aircraft, it seems that the evolution of airplane types has converged to what is sometimes called the "wing-and-tube" configuration. Differences seem to be marginal and only visible to the experts. From far, an Airbus A350 XWB does not look that different with respect to the first Boeing

Dash-80 (later B707) – except for the number of engines. But a lot has happened inside the wing and especially at the high-lift system.

Peter K.C. Rudolph analyzed in detail the development of high-lift system technologies of the flying aircraft in 1996 [16]. His introduction is a very good summary of the development path until then:

> The early breed of slow commercial airliners did not require high-lift systems because their wing loadings were low and their speed ratios between cruise and low speed (takeoff and landing) were about 2:1. However, even in those days the benefit of high-lift devices was recognized. Simple trailing-edge flaps were in use, not so much to reduce landing speeds, but to provide better glideslope control without sideslipping the airplane and to improve pilot vision over the nose by reducing attitude during low-speed flight.
>
> As commercial-airplane cruise speeds increased with the development of more powerful engines, wing loadings increased and a real need for high-lift devices emerged to keep takeoff and landing speeds within reasonable limits. The high-lift devices of that era were generally trailing-edge flaps. When jet engines matured sufficiently in military service and were introduced commercially, airplane speed capability had to be increased to best take advantage of jet engine characteristics. This speed increase was accomplished by introducing the wing sweep and by further increasing wing loading. Whereas increased wing loading called for higher lift coefficients at low speeds, wing sweep actually decreased wing lift at low speeds.
>
> Takeoff and landing speeds increased on early jet airplanes, and, as a consequence, runways worldwide had to be lengthened. There are economical limits to the length of runways; there are safety limits to takeoff and landing speeds; and there are speed limits for tires. So, in order to hold takeoff and landing speeds within reasonable limits, more powerful high-lift devices were required. Wing trailing-edge devices evolved from plain flaps to Fowler flaps with single, double, and even triple slots. Wing leading edges evolved from fixed leading edges to a simple Krueger flap, and from fixed, slotted leading edges to two- and three-position slats and variable-camber (VC) Krueger flaps.
>
> The complexity of high-lift systems probably peaked on the Boeing 747, which has a VC Krueger flap and triple-slotted, inboard and outboard trailing-edge flaps. Since then, the tendency in high-lift system development has been to achieve high levels of lift with simpler devices in order to reduce fleet acquisition and maintenance costs.

**(Rudolph [16] p. 1)**

In fact, especially the Airbus A320 demonstrated a low level of complexity that was not believed to be sufficient for large transport aircraft cruising at high subsonic speeds. It was the first aircraft of its class relying completely on a single-slotted Fowler flap only. Rudolph states on this:

> It has long been believed that the single-slotted flap is not powerful enough to provide an acceptable landing attitude; however, the recently developed Airbus models A320, A330, and A340 prove otherwise.

**(Rudolph [16] p. 137)**

The development past the mid-90s has focused even more on simplifying the system to achieve benefits regarding less system weight and maintenance effort. Looking again at the two most modern aircraft today – the Boeing B787 Dreamliner and the Airbus A350 XWB – similar high-lift technology is applied: They use slats at the leading edge (at the A350 XWB additionally replaced by a droop nose inside the engine) and single slotted flaps on fixed hinges in combination with adjustable spoilers to control the flap gap – a technology developed by Douglas during the YC-15/C-17 program and strongly suggested by Rudolph in his summary for further development.

## 1.2   THE LIMITS OF SAFE FLIGHT

From the beginning, the major aim of high-lift devices has been to reduce the landing speed to make landing more controllable and safer. For example, Gustav Lachmann started his work on slats due to an accident he had during landing. In consequence, high-lift systems are intended to extend the limits of safe flight towards lower flight speeds.

Before going into details, two speed definitions are needed that are incorporated into the speed limits. The True Airspeed (TAS) is the real speed of the aircraft with respect to the still air, or by relativity theory, the real flow speeds the aircraft experiences. The Equivalent Airspeed (EAS) is the airspeed that the aircraft would experience at sea level in standard atmosphere resulting in the same dynamic pressure. It is especially important for structural considerations as the aerodynamic forces on an aircraft scale with the dynamic pressure. The relation of TAS and EAS is given by the density ratio:

$$EAS = TAS \sqrt{\frac{\rho}{\rho_{H=0}}}. \tag{1.3}$$

The limits of flight can be expressed by the envelope of conditions where the aircraft can be operated safely. There are two types of envelopes that visualize the allowable conditions of flight. The first visualization is the V-n-diagram depicted in Figure 1.6. It shows the allowable maneuver load factor over the flight speed. The lower flight limit is given by the stall speed without maneuver load $V_{S1g}$ that corresponds to the aerodynamic maximum lift coefficient. As no higher lift coefficient can be achieved, positive maneuver loads can only be compensated at higher speeds. By this, this is an **aerodynamic limit**. A similar boundary is given by the minimum lift coefficient in the range of negative load factors. As explained later on, when regarding certification regulations, a transport aircraft has to be designed for positive maneuver load factors up to $n_z = 2.5$ and down to $n_z = -1$. Further on, for an aircraft, a speed is defined, in which the aircraft can be flown safely even in gusty conditions, the so-called maximum operating speed $V_{MO}$. This speed limit can be released slightly if in this case maneuvers are restricted to reduced load factors, but even then, a certain speed – the never exceed speed $V_{NE}$ – has to be respected to not overload the aircraft. These limits are **restrictions from structural considerations**. It has to be noted, that the V-n-diagram is not constant for an aircraft. As the stall speed is linked to the

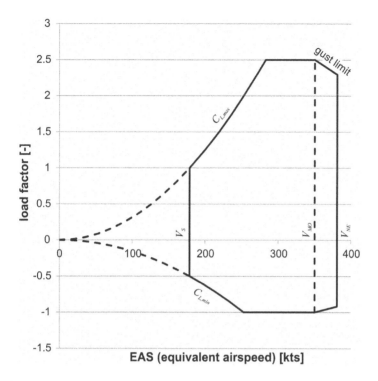

**FIGURE 1.6** V-n-diagram of allowable load limits related to the flight speed.

maximum lift coefficient and the aircraft mass, the lower bound is affected by the actual aircraft weight. The upper bound is also variable when flying at high altitudes, but this can be better expressed by the second type of envelope.

The second visualization is the V-H-diagram in Figure 1.7 showing the allowed flight speed vs. the altitude. It describes the variation of the flight speed limits at steady flight ($n_z = 1$) for the variation of the altitude and is in this case shown for the TAS. Therefore, the airspeed increases with altitude to compensate for the reduction of air density. At lower altitudes, the limits for minimum and maximum speed are the same as in the V-n-diagram. At higher altitudes a different boundary appears, which is not visible in the V-n-diagram. As the temperature decreases, the speed of sound decreases, too, as they are linked together for an ideal gas by:

$$a = \sqrt{\kappa R T}. \tag{1.4}$$

So, with increasing altitude, the aircraft has to fly faster to keep the same stagnation pressure, but it approaches the sonic speed more closely. As transonic effects, and especially the so-called buffeting – an unstable interaction of supersonic shocks and thereof induced flow separation – may appear, in addition to the maximum operating and never exceed speeds, similar boundaries have to be specified in terms of Mach numbers ($M_{MO}$, $M_{NE}$) to safely stay off this **aerodynamic limitation**. The graph

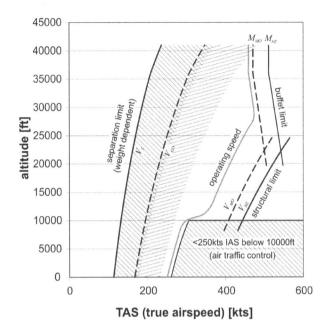

**FIGURE 1.7**   V-H-diagram of flight speed limits related to flight altitude.

shown already includes the variation of the stall speed with the aircraft weight. It additionally shows two further characteristics. The first is the so-called green dot speed $V_{GD}$. This is the speed of the best lift-to-drag ratio. The name is derived from the actual aircraft avionics that displays this condition by a green dot at the speed indicator. As there is no reason to fly an aircraft in cruise configuration with less speed than the $V_{GD}$, this is the natural lowest flight speed in operation. The second is the target operating speed, shown in the graph too, which is the optimum condition regarding the range. It is closer to the maximum operating limits.

An additional boundary shown in the V-H-diagram is related to **air traffic control (ATC) considerations**. At an altitude below FL100 (flight level 10,000 ft above 1013.25 hPa in standard atmosphere), the civil transport aircraft traffic merges with other types of flight activities, especially the general aviation. To increase safety, the speed differences between different aircraft are restricted by regulations through air traffic control regulations [17] in most countries to limit the maximum EAS to 250 knots. As the V-H-diagram shows for the unnamed exemplary aircraft, this speed restriction is close to – or can already be less than – the minimum rational flight speed at high aircraft weights.

The major functionality of high-lift systems is to reduce the minimum flight speed. To achieve this, the shown envelopes have to be extended into the direction of lower speeds. High-lift devices achieve this by shifting the aerodynamic boundary associated with the maximum lift coefficients to lower speeds. Figure 1.8 shows the V-n-diagram including a high-lift system with two distinct deflections (e.g., for take-off and landing). It can be expected that high-lift systems are used in a stage of flight

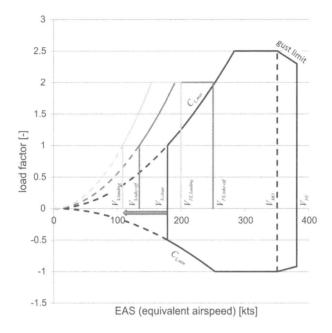

**FIGURE 1.8** Intended changes of the operational envelope by high-lift systems in the V-n-diagram.

where high maneuverability is less important. The maximum required maneuver load factor $n_z = 2$ is lower than for cruising configurations, and no negative maneuver load is expected.

Designing a high-lift system for the full flight envelope with regard to high speeds would make the supporting structures unnecessarily heavy, as the device will not be in use in these conditions. Therefore, the structural limit for high-lift systems is significantly lower. The major requirement is of course that the envelopes of the different settings overlap with a sufficient speed flexibility for the transition process.

Figure 1.9 shows the V-H-diagram with the high-lift system included. For simplicity, the clean wing configuration is reduced to its operational limits. Again, only two distinct high-lift device deflections are shown. While the corresponding stall speeds ($V_{S0}$ for landing, $V_{S2}$ for take-off) show the limit capabilities, there has to be a safety margin for the minimum flight speed that should not be underrun in operations. These two important limits are the minimum approach speed $V_{REF}$ and the minimum climb speed $V_2$ that will be further discussed when looking at the flight safety regulations in Chapter 3. Also from these regulations, the maximum speed with deflected high-lift devices $V_{FE}$ has to be specified by the aircraft manufacturer for each high-lift setting separately. Additionally, it can be seen, that the envelope for a high-lift system does not target the whole altitude range. A common limit for the maximum altitude is seen at FL200 (flight level 20,000 ft above 101325 Pa in standard atmosphere).

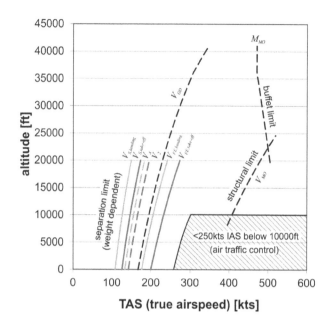

**FIGURE 1.9** V-H-diagram including high-lift devices; for simplicity only the operational envelope of the clean wing configuration is shown.

## NOTE

1. US government shut down public access to its Digital Aeronautical Flight Information File (DAFIF) service in 2006 due to copyright issues. OurAirports.com is a public domain initiative to fill this gap of available airport data for research.

## REFERENCES

[1] OurAirports (2007) Open Data Downloads, accessed April 2021 @ https://ourairports.com/data/airports.csv and https://ourairports.com/data/runways.csv.
[2] Prandtl L (1904) "Über die Flüssigkeitsbewegung bei sehr kleiner Reibung," III Internationaler Mathematiker-Kongress, Heidelberg, translated to English as NACA TM 452.
[3] Nayler, Stedman, Stern (1914) "Experiments on an Aerofoil Having a Hinged Rear Portion," ARC R&M No. 110.
[4] Lachmann G (1921) Das unterteilte Flächenprofil, Zeitschrift für Flugtechnik und Motorluftschiffahrt **12**, pp. 164–169, translated to English as: NACA TN 71.
[5] Handley Page F (1920) Wing and Similar Member of Aircraft, patent US 1,353,666.
[6] Fowler HD (1921) Variable Area Wing, patent US 1,392,005.
[7] Fowler HD (1928) Aerofoil, patent US 1,670,852.
[8] Bölkow L (1942) Tragflügel mit Mitteln zur Veränderung der Profileigenschaften, patent DE 694,916.
[9] Krueger W (1943) Über eine neue Möglichkeit der Steigerung des Höchstauftriebes von Hochgeschwindigkeitsprofilen, AVA-report 43/W/64, Aerodynamische Versuchsanstalt Göttingen.
[10] Watts FT (1968) Aircraft, patent US 3,363,859.

[11] Krueger W (1943) Hochauftrieb. Zusammenstellung und Vergleich verschiedener Bauarten und Methoden, AVA-report 43/W/38, Aerodynamische Versuchsanstalt Göttingen.

[12] Weyl AR (1945) "High-Lift Devices and Tailless Airplanes," Aircraft Engineering, **17**(10), pp. 292–297.

[13] Smith AMO (1975) "High-Lift Aerodynamics," Journal of Aircraft, **12**(6), pp. 501–530. DOI: 10.2514/3.59830.

[14] Betz A (1922) Theory of the Slotted Wing, NACA TN 100. Translation of Berichte und Abhandlungen der wissenschaftlichen Gesellschaft für Luftfahrt (Supplement to Zeitschrift für Flugtechnik und Motorluftschiffahrt), No. 6, January, 1922.

[15] Pleines W (1935) Wing Brake Flaps, Aircraft Engineering, **7**(9), pp. 213–219.

[16] Rudolph PKC (1993) High-Lift Systems on Commercial Subsonic Airliners, NASA CR 4746.

[17] Code of Federal Regulations Title 14 Part 91 §91.117 (1993) Amdt. 91–233, 58 FR 43554.

# 2 Limits of Lift Generation

## NOMENCLATURE

| Symbol | Unit | Description | Symbol | Unit | Description |
|---|---|---|---|---|---|
| $a$ | $m/s$ | Speed of sound | $\beta$ | – | Falkner-Skan coefficient |
| $b$ | $m$ | Span | $\gamma$ | $°(deg)$ | Flight path angle |
| $c$ | $m$ | Chord length | $\Gamma$ | $m^2/s$ | Circulation |
| $c_f$ | – | Friction coefficient | $\delta$ | $m$ | Displacement thickness |
| $c_p$ | – | Pressure coefficient | $\delta_N$ | $m$ | Characteristic boundary layer thickness |
| $c_p^*$ | – | Critical pressure coefficient | $\zeta$ | $m$ | Transformed complex coordinate (conformal mapping) |
| $\bar{c}_p$ | – | Canonical pressure coefficient | $\eta,\xi$ | $m$ | Transformed coordinates |
| $c_{L_a}$ | – | Upper side contribution to sectional lift coefficient | $\theta$ | $m$ | Momentum loss thickness |
| $C_L$ | – | Lift coefficient | $\kappa$ | – | Ratio of specific heat coefficients (1.4 for ideal gas) |
| $E$ | $J/kg$ | Specific internal energy | $\lambda$ | $W/m^2$ | Heat flux coefficient |
| $f$ | $m^2/s$ | Wall normal portion of the stream function | $\mu$ | $m^2/s$ | Dipole strength (potential theory) |
| $F$ | – | Complex flow function | $\mu$ | $kg/m\ s$ | Dynamic viscosity |
| $H$ | $J/kg$ | Specific enthalpy | $\nu$ | $m^2/s$ | Kinematic viscosity |
| $H_{12}$ | – | Form factor | $\rho$ | $kg/m^3$ | Density |
| $k$ | – | Constant | $\sigma$ | $m^2/s$ | Source emissivity (potential theory) |
| $l$ | $m$ | Characteristic length | $\tau$ | $N/m^2$ | Viscous stress tensor |
| $L$ | $N$ | Lift force | $\tau_w$ | $N/m^2$ | Wall shear stress |
| $m$ | – | Exponent | $\Phi$ | $m^2/s$ | Potential |
| $M$ | – | Mach number | $\Psi$ | $m^2/s$ | Stream function |
| $\mathbf{n}$ | $m$ | Surface normal vector pointing into the flow domain | $\omega$ | $m/s$ | Transformed complex velocity (conformal mapping) |
| $p$ | $N/m^2$ | Pressure | $\Omega$ | $m^2$ | Body surface |
| $p_t$ | $N/m^2$ | Stagnation pressure | $\Delta$ | | Difference operator |
| $R$ | $J/kg\ K$ | Specific gas constant | $\nabla$ | | Gradient operator |
| $R$ | $m$ | Cylinder radius | $\otimes$ | | Outer product operator |
| $r$ | $m$ | Radius | $\infty$ | | Onflow condition |
| $Re$ | – | Reynolds number | c | | Center |
| $S$ | – | Stratford criterion | c | | Compressible |
| $t$ | $s$ | Time | e | | Boundary layer edge |
| $T$ | $K$ | Temperature | ic | | Incompressible |
| $\mathbf{u}$ | $m/s$ | Velocity vector | vac | | Vacuum |
| $u,v,w$ | $m/s$ | Components of velocity vector | max | | Maximum value |

DOI: 10.1201/9781003220459-2

| | | | | | |
|---|---|---|---|---|---|
| $U$ | $m/s$ | Flow velocity | min | | Minimum value |
| $V$ | $m^3$ | Volume | $r$ | | Radial |
| $w$ | $m/s$ | Complex velocity | $D$ | | Dipole |
| $\bar{w}$ | $m/s$ | Conjugate complex velocity | $s$ | | Source/sink |
| $x, y, z$ | $m$ | Cartesian coordinates | $\theta$ | | Circumferential |
| $z$ | $m$ | Complex coordinate | $\Gamma$ | | Potential vortex |
| $\alpha$ | $°(deg)$ | Angle of attack | $\parallel$ | | Parallel flow |
| $\beta$ | $°(deg)$ | Opening angle | $0$ | | Start of pressure rise |

Before analyzing the physical limits of lift generation, it is necessary to recapitulate the basic mathematical description of fluid flows[1]. In aerodynamics, air is generally assumed to be an ideal gas with the corresponding material properties. The motion of air is described by the Navier-Stokes equations that formulate the principles of mass, momentum, and energy conservation for a Newtonian fluid.

$$\frac{D}{Dt}\iiint_V \rho\, dV = \iiint_V \frac{\partial \rho}{\partial t}\, dV + \iint_\Omega \rho(\mathbf{u}\cdot\mathbf{n})d\Omega = 0$$

$$\frac{D}{Dt}\iiint_V \rho\mathbf{u}\, dV = \iiint_V \frac{\partial(\rho\mathbf{u})}{\partial t}\, dV + \iint_\Omega \rho\mathbf{u}(\mathbf{u}\cdot\mathbf{n})d\Omega = -\iint_\Omega p\mathbf{n}\, d\Omega + \iint_\Omega (\tau\cdot\mathbf{n})d\Omega$$

$$\frac{D}{Dt}\iiint_V \rho E\, dV = \iiint_V \frac{\partial(\rho E)}{\partial t}\, dV + \iint_\Omega \rho H(\mathbf{u}\cdot\mathbf{n})d\Omega = \iint_\Omega \lambda(\nabla T\cdot\mathbf{n})d\Omega + \iint_\Omega (\tau\cdot\mathbf{n})\cdot\mathbf{u}\, d\Omega.$$

$$(2.1)$$

The set of equations is closed by the universal gas law

$$\frac{p}{\rho} = RT. \tag{2.2}$$

For subsonic flows without discontinuities, the Navier-Stokes equations can be transformed into differential form using the Gauss theorem

$$\frac{\partial \rho}{\partial t} + \nabla\cdot(\rho\mathbf{u}) = 0$$

$$\frac{\partial(\rho\mathbf{u})}{\partial t} + \nabla\cdot(\rho\mathbf{u}\otimes\mathbf{u}) = -\nabla p + \nabla\cdot\tau \tag{2.3}$$

$$\frac{\partial(\rho E)}{\partial t} + \nabla\cdot(\rho H\mathbf{u}) = \nabla\cdot(\lambda\nabla T) + \nabla\cdot(\tau\cdot\mathbf{u}).$$

Dimensional analysis results in two important scaling numbers, the Mach number

$$M = \frac{U_\infty}{a_\infty}; \; a_\infty = \sqrt{\kappa R T_\infty} \tag{2.4}$$

and the Reynolds number

$$\mathrm{Re}_l = \frac{\rho_\infty U_\infty l}{\mu}. \tag{2.5}$$

Two flows are similar only if these two numbers are equal.

When assuming inviscid flow by setting the viscosity to zero, the stress tensor $\tau$ vanishes and both forms result in the Euler equations. The integration of the momentum equation along a streamline results in the Bernoulli equation

$$p_t = p + \frac{1}{2}\rho\|\mathbf{u}\|^2 = p_\infty + \frac{1}{2}\rho_\infty U_\infty^2 = const. \tag{2.6}$$

relating the local flow speed to the local pressure. For dimensionless analysis, the local pressure coefficient is defined by

$$c_p = \frac{p - p_\infty}{p_{t,\infty} - p_\infty} = \frac{p - p_\infty}{\frac{1}{2}\rho_\infty U_\infty^2}. \tag{2.7}$$

While the above form is suited to be evaluated directly in incompressible flow, in compressible flow the variation of the density requires additionally taking into account the gas law. Assuming adiabatic changes of the fluid state along the streamline leads to the formulation

$$p_t = p\left(1 + \frac{\kappa - 1}{2}M^2\right)^{\frac{\kappa}{\kappa-1}} = p_\infty\left(1 + \frac{\kappa - 1}{2}M_\infty^2\right)^{\frac{\kappa}{\kappa-1}} = const. \tag{2.8}$$

Additionally, regarding the flow as incompressible simplifies again the set of equations to

$$\nabla \cdot \mathbf{u} = 0$$

$$\rho\frac{d\mathbf{u}}{dt} + \rho\nabla\cdot(\mathbf{u}\otimes\mathbf{u}) = -\nabla p \tag{2.9}$$

$$\frac{d\rho E}{dt} + \rho\nabla\cdot H\mathbf{u} = \nabla\cdot(\lambda\nabla T).$$

A vector field having no divergence and no rotation

$$\nabla\times\mathbf{u} = \mathbf{0} \tag{2.10}$$

is a potential field. The potential theory in aerodynamics describes therefore the solution of the incompressible conservation law by using a potential field. For fluid dynamics, a potential $\Phi$ is defined so that the spatial derivatives are the velocity components

$$\Phi_x = \frac{\partial\Phi}{\partial x} = u$$

$$\Phi_y = \frac{\partial\Phi}{\partial x} = v. \tag{2.11}$$

Using this in the incompressible mass conservation equation leads to the linearized potential equation for fluid dynamics[2]

$$\Phi_{xx} + \Phi_{yy} = 0 \tag{2.12}$$

For each potential field $\Phi$, a stream function $\Psi$ is defined that is orthogonal to the potential field in every point

$$\Psi_x = \Phi_y$$
$$\Psi_y = -\Phi_x. \tag{2.13}$$

The important property of potential fields is that the summation of two potential fields results in another potential field, enabling the principle of superposition.

The three basic solutions for eq. (2.12) within fluid dynamics, which are shown in Figure 2.1, are

1. the parallel flow

$$\Phi_{\parallel}(x,y) = U_\infty\left(x\cos\alpha + y\sin\alpha\right)$$
$$u(x,y) = U_\infty\cos\alpha \tag{2.14}$$
$$v(x,y) = U_\infty\sin\alpha;$$

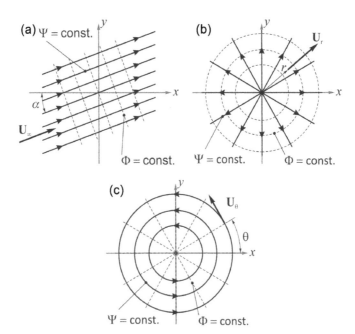

**FIGURE 2.1**   Basic singular solutions for the potential flow field equation: (a) parallel flow; (b) source/sink; (c) potential vortex.

2. the source/sink flow with emissivity $\sigma$ inducing the radial velocity $\mathbf{U}_r$

$$\Phi_S(x,y) = \frac{\sigma}{2\pi} \ln \sqrt{x^2 + y^2}$$

$$u(x,y) = \frac{\sigma}{2\pi} \frac{x}{x^2 + y^2} \tag{2.15}$$

$$v(x,y) = \frac{\sigma}{2\pi} \frac{y}{x^2 + y^2};$$

3. the potential vortex with circulation $\Gamma$ inducing the circumferential velocity $\mathbf{U}_\theta$

$$\Phi_\Gamma(x,y) = \frac{\Gamma}{2\pi} \arctan\left(\frac{y}{x}\right)$$

$$u(x,y) = -\frac{\Gamma}{2\pi} \frac{y}{x^2 + y^2} \tag{2.16}$$

$$v(x,y) = \frac{\Gamma}{2\pi} \frac{x}{x^2 + y^2}.$$

Figure 2.2 shows a fourth elementary solution obtained by superposition of a source and a sink at the same position. It is the dipole of strength $\mu$

$$\Phi_D(x,y) = -\frac{\mu}{2\pi} \frac{x}{x^2 + y^2}$$

$$u(x,y) = \frac{\mu}{2\pi} \frac{x^2 - y^2}{\left(x^2 + y^2\right)^2} \tag{2.17}$$

$$v(x,y) = \frac{\mu}{2\pi} \frac{2xy}{\left(x^2 + y^2\right)^2}.$$

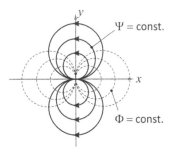

**FIGURE 2.2**   Potential field of a dipole.

## 2.1  INCOMPRESSIBLE, INVISCID FLOWS

To approach the limits of lift generation, first, the flow simplest to analyze is regarded. This will justify, whether even in incompressible and inviscid flow a theoretical limit for lift generation can be derived. For this, the first flow makes use of the potential theory.

The superposition of the parallel flow, a dipole, and a potential vortex leads to a flow field shown in Figure 2.3. It represents the flow around a "rotating" cylinder in horizontal onflow ($\alpha = 0$). The corresponding potential field is given by

$$\Phi(x,y) = U_\infty x - \frac{\mu}{2\pi}\frac{x}{x^2+y^2} + \frac{\Gamma}{2\pi}\arctan\left(\frac{y}{x}\right)$$

$$u(x,y) = \Phi_x(x,y) = U_\infty + \frac{\mu}{2\pi}\left(\frac{x^2-y^2}{\left(x^2+y^2\right)^2}\right) - \frac{\Gamma}{2\pi}\left[\frac{y}{x^2+y^2}\right] \qquad (2.18)$$

$$v(x,y) = \Phi_y(x,y) = \frac{\mu}{2\pi}\frac{2xy}{\left(x^2+y^2\right)^2} + \frac{\Gamma}{2\pi}\left[\frac{x}{x^2+y^2}\right].$$

Analyzing this flow field, the diameter of the cylinder is directly linked to the dipole strength. But what limits the generation of lift force for this simple problem? What is seen in the flow field are the two distinct stagnation points. Their positions on the cylinder surface are directly related to the circulation. The question, therefore, is, if a solution exists where these two stagnation points meet each other on the lower edge of the cylinder, as this is obviously the situation where the maximum lift is produced with having a stagnation point on the surface.

Now, looking at the flow field shown in Figure 2.4 leads to the side constraints to be fulfilled by the potential field eq. (2.18) assuming the radius to be 1 to eliminate dimensionalities

$$u(0;-1) = v(0;-1) = u(-1;0) = 0$$
$$v = 0 \;\forall\; x = 0. \qquad (2.19)$$

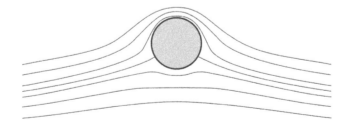

**FIGURE 2.3**  Flow field around combination of parallel flow, dipole, and potential vortex. (Reproduction of figure 9(a) from Smith [5].)

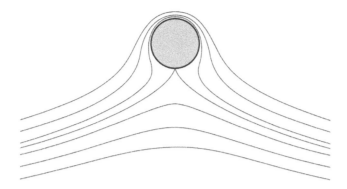

**FIGURE 2.4** Limit flow around cylinder, where force and aft stagnation points align. (Reproduction of figure 9(b) from Smith [5].)

While the condition along the center axis defines the dipole strength, the position of the stagnation point defines the circulation as expected

$$u(-1;0) = U_\infty + \frac{\mu}{2\pi} = 0 \Rightarrow \mu = -2\pi U_\infty$$

$$u(0;-1) = U_\infty - \frac{\mu}{2\pi} + \frac{\Gamma}{2\pi} = 0 \Rightarrow \Gamma = -2\pi U_\infty + \mu = -4\pi U_\infty. \tag{2.20}$$

Using Zhukovsky's theorem for the relation of circulation and lift force

$$L = -\rho_\infty \Gamma b U_\infty \tag{2.21}$$

and the definition of the lift coefficient

$$C_L = \frac{L}{\frac{1}{2}\rho_\infty U_\infty^2 bc} \tag{2.22}$$

leads – keeping in mind that $c = 2r = 2$ – to the lift coefficient corresponding to this limit flow of

$$C_{L,max} = 4\pi. \tag{2.23}$$

So, this is the limiting maximum lift coefficient for an incompressible inviscid flow, where a stagnation point on the object is present.

One may argue that the presented idea of a rotating cylinder is somehow artificial. The rotating cylinder would need friction to impose its rotation on the flow. So, it is not consequent to assume inviscid flow in this scenario. But this elementary solution seen above serves as another basis to approach airfoil flows analytically by potential theory.

To introduce this, first, a transformation is introduced from the Cartesian space into a complex space by

$$z = x + iy$$
$$w = u + iv. \qquad (2.24)$$

As this is a direct coordinate transformation, the potential is unchanged

$$\Phi(z) = \Phi(x, y). \qquad (2.25)$$

To obtain the basic solutions from eqs. (2.14) to (2.17) of the potential field in complex space, first the complex flow function is defined as

$$F(z) = \Phi + i\Psi. \qquad (2.26)$$

The velocity is obtained from the flow function by the conjugate complex

$$\frac{dF}{dz} = u - iv = \bar{w}. \qquad (2.27)$$

The complex flow functions of the basic solutions are then, for the singularity, placed at $z_c$

$$F_\| = \bar{w}_\infty \cdot z; \qquad (2.28)$$

$$F_S = \frac{\sigma}{2\pi} \ln(z - z_c); \qquad (2.29)$$

$$F_\Gamma = i \frac{\Gamma}{2\pi} \ln(z - z_c); \qquad (2.30)$$

$$F_D = \frac{\mu}{2\pi} \frac{1}{(z - z_c)}. \qquad (2.31)$$

From the theory of conformal mapping, it is known that these transformations are isogonal. So, applying a conformal mapping to a potential field results again in a potential field, as the orthogonality between potential and stream function is maintained.

The transformation of Kutta and Zhukovsky

$$\zeta = z + \frac{1}{z} \qquad (2.32)$$

is such a conformal mapping. It transforms the unit circle to a flat line of length 4, see Figure 2.5. Furthermore, this mapping transforms a circle passing the point $z = 1$ into an airfoil shape with a sharp trailing edge depending on the center point of the circle.

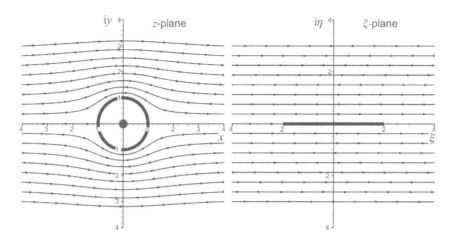

**FIGURE 2.5**  Kutta-Zhukovsky transformation of the potential field around a circle into the flow field around a flat plate.

Figure 2.6 shows the results of the transformation when shifting the center point of the circle. Moving the center along the negative real axis introduces thickness while moving the center up along the positive imaginary axis introduces camber. An important point now is to remember that the $z = 1$ in the $z$-plane is mapped to the sharp trailing edge $\zeta = 2$ in the mapped $\zeta$-plane. Introducing the Kutta-condition[3], which is nothing more than to fix the rear stagnation point at the sharp trailing edge in the $\zeta$-plane, or the point $z = 1$ respectively, gives the unique solution for the flow field in the $z$-plane depending on the center point, the flow speed, and the angle of attack by

$$R = |1 - z_c|$$
$$\mu = 2\pi R^2 U_\infty \tag{2.33}$$
$$\Gamma = 4\pi R U_\infty \sin(\alpha + \beta),$$

where $\beta$ is the angle between the line from the trailing edge to the center point and the real axis, we call it the opening angle of the airfoil (Figure 2.7). The corresponding flow field is then evaluated to

$$w(z) = U_\infty e^{-i\alpha} + \frac{i\Gamma}{2\pi(z - z_c)} - U_\infty e^{i\alpha} \frac{R^2}{(z - z_c)^2}. \tag{2.34}$$

In the $\zeta$-plane the mapped velocity is

$$\omega(\zeta) = \frac{w(z)}{1 - \dfrac{1}{z^2}}. \tag{2.35}$$

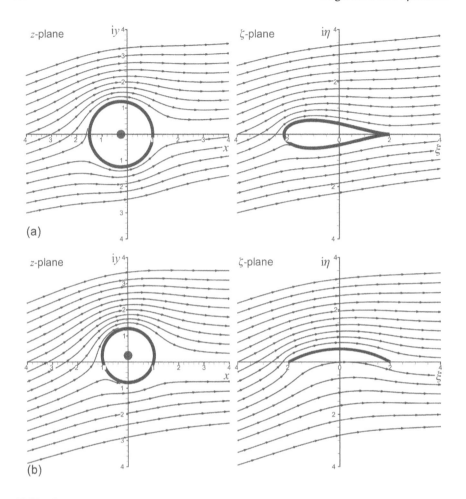

**FIGURE 2.6** Variation of mapped shape by movement of the center point of Kutta-Zhukovsky transform: (a) thickness by movement to negative real axis; (b) camber by movement to positive imaginary axis.

Neglecting the effect of thickness, the movement of the center point along the positive imaginary axis leads to camber lines of arc shape. The lift coefficient is then given by

$$
C_L = \begin{cases} 2\pi \sin(\alpha+\beta)/\cos(\beta) & \text{for} \quad 0° \le \beta \le 45° \\ 4\pi \sin(\beta)\sin(\alpha+\beta) & \text{for} \quad 45° < \beta \le 90° \end{cases} \tag{2.36}
$$

It is easily seen that the maximum lift coefficient is obtained in any case when the sum of the angle of attack and the opening angle of the airfoil reaches 90°. For different camber lines, the corresponding maximum lift coefficients are summarized in Table 2.1.

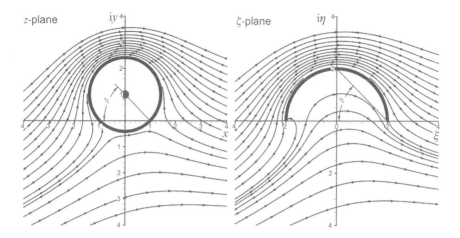

**FIGURE 2.7** Definition of opening angle for evaluation of the Kutta-Zhukowsky transform.

Multi-element airfoils can achieve camber lines comparable to quarter arcs. Practical experience with multi-element airfoils showed that maximum lift coefficients above $C_{L,\max} \approx 5$ are achievable.

On the one hand, this indicates that the potential flow solutions are close, but on the other hand, that additional effects limit the generation of lift. They have to be linked to the influences neglected in potential theory, which are compressibility and viscosity.

## 2.2 COMPRESSIBILITY EFFECTS

In general, air is a compressible fluid. A common assumption for incompressible flow is derived from the relation when the effect of compressibility gets significant. One of these approaches compares the linear potential eq. (2.12) with the compressible potential equation related to small disturbance theory by Prandtl and Glauert

$$\left(1 - M_\infty^2\right)\Phi_{xx} + \Phi_{yy} = 0. \tag{2.37}$$

### TABLE 2.1
### Maximum Lift Coefficients of Arc Camber Lines in Incompressible, Inviscid Flow according to Potential Theory by Kutta-Zhukovsky Transform

| Arc Type | Opening Angle β | Angle of Attack α | Maximum Lift Coefficient $C_L$ |
|---|---|---|---|
| Flat plate | 0° | 90° | $2\pi \approx 6.28$ |
| Quarter arc | 22.5° | 67.5° | $\dfrac{2\pi}{\cos(22.5°)} \approx 6.8$ |
| Half arc | 45° | 45° | $\dfrac{4\pi}{\sqrt{2}} \approx 8.9$ |
| Full arc | 90° | 0° | $4\pi \approx 12.56$ |

From the comparison of both, the Prandtl-Glauert correction for thin airfoils is derived for the ratio of the pressure coefficients in compressible and incompressible flow

$$\frac{c_{p,\mathrm{ic}}}{c_{p,\mathrm{c}}} = \sqrt{1 - M_\infty^2}.$$

(2.38)

The commonly used limit for incompressible flow is derived from the relation when the introduced error exceeds 5%, leading to the limiting Mach number of

$$\sqrt{1 - M_\infty^2} < 0.95 \quad \Rightarrow \quad M_\infty > \sqrt{1 - 0.95^2} \approx 0.31.$$

(2.39)

We will see later whether this assumption is valid to exclude the relevance of compressibility at extreme lift conditions.

First, it is worth to take a look where compressibility limits the generation of lift directly. One obvious limit due to compressibility is that, in a compressible medium, the pressure by definition cannot be less than $p = 0$, as this value is corresponding to a vacuum. Using the definition of the pressure coefficient in compressible flow

$$c_p = \frac{p - p_\infty}{\frac{1}{2} \kappa p_\infty M_\infty^2}$$

(2.40)

and setting the actual pressure to vacuum condition leads to the absolute limit condition for the minimum pressure coefficient depending on the onflow Mach number

$$c_{p,\mathrm{vac}} M_\infty^2 = -\frac{2}{\kappa} \approx -1.4.$$

(2.41)

Experiments by Mayer [6] showed that vacuum is not achieved with attached flow. Figure 2.8 summarizes the analysis by Mayer. From available experimental data, the

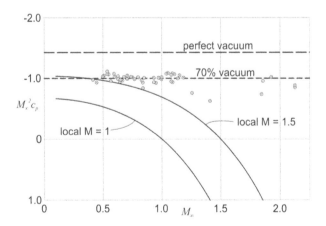

**FIGURE 2.8** Minimum achievable pressure coefficient at attached flow according to experimental data analyzed by Mayer [6]. (Reproduction of figure 12 from Smith [5][4].)

minimum value of $c_p M_\infty^2$ is plotted versus the actual Mach number. The data at very different Mach numbers showed a practical limit at $c_p M_\infty^2$, which corresponds to 70% vacuum. Therefore, the **minimum pressure coefficient** is given by the Mayer-limit

$$c_{p,\min} M_\infty^2 \geq -1. \tag{2.42}$$

The next important condition in compressible flow is when the flow changes its characteristics from subsonic to supersonic flow. This is given by the critical pressure coefficient, where the local flow speed reaches the speed of sound. Along a streamtube, the compressible Bernoulli eq. (2.8) is valid. The pressure ratio depending on the local Mach number is therefore given by

$$\frac{p}{p_\infty} = \frac{\left(1 + \dfrac{\kappa - 1}{2} M_\infty^2\right)^{\frac{\kappa}{\kappa-1}}}{\left(1 + \dfrac{\kappa - 1}{2} M^2\right)^{\frac{\kappa}{\kappa-1}}}. \tag{2.43}$$

Setting the sonic condition leads to the critical pressure coefficient

$$c_p^* = \frac{\left(1 + \dfrac{\kappa - 1}{2} M_\infty^2\right)^{\frac{\kappa}{\kappa-1}} - \left(1 + \dfrac{\kappa - 1}{2}\right)^{\frac{\kappa}{\kappa-1}}}{\dfrac{\kappa}{2} M_\infty^2 \left(1 + \dfrac{\kappa - 1}{2}\right)^{\frac{\kappa}{\kappa-1}}}. \tag{2.44}$$

Figure 2.9 shows the evolution of the critical pressure coefficient and the Mayer limit for the range of subsonic flow speeds. The magnitudes grow for low Mach numbers

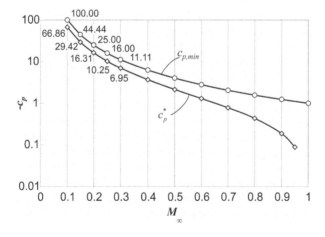

**FIGURE 2.9**  Characteristic values of the Mayer limit and the critical pressure coefficient for various subsonic flow Mach numbers.

but are not extremely high. Remembering that, in incompressible flow, the pressure coefficient is directly related to the velocity ratio

$$c_{p,ic} = 1 - \left( \frac{|\mathbf{u}|}{U_\infty} \right)^2 \tag{2.45}$$

e.g., a pressure coefficient of $c_{p,ic} = -24$ relates to a local speed five times the onflow speed. At high angles of attack having a high acceleration of the flow around the leading edge, this is not out of imagination.

Coming back to the common assumption of incompressible flow for $M_\infty \le 0.3$ derived from the Prandtl-Glauert correction eq. (2.38), it needs to be revisited whether high-lift flows are covered by the small disturbance theory of thin airfoils. The Prandtl-Glauert correction gives a relation of the local pressure coefficient based on one single global parameter of the flow: the Mach number of the free stream. But the corresponding error is local. Therefore, improved compressibility corrections have been derived by approximating the compressible pressure coefficient by relating to the local Mach number. Two formulations are commonly used, the Kármán-Tsien Rule [7]

$$c_{p,c} = \frac{c_{p,ic}}{\sqrt{1 - M_\infty^2} + \dfrac{M_\infty^2}{1 + \sqrt{1 - M_\infty^2}} \dfrac{c_{p,ic}}{2}} \tag{2.46}$$

or the correction by Laitone [8]

$$c_{p,c} = \frac{c_{p,ic}}{\sqrt{1 - M_\infty^2} + \dfrac{M_\infty^2}{1 + \sqrt{1 - M_\infty^2}} \left( 1 + \dfrac{\kappa - 1}{2} M_\infty^2 \right) \dfrac{c_{p,ic}}{2}}. \tag{2.47}$$

Figure 2.10 depicts the limiting relation when these corrections exceed 5% difference between compressible and incompressible pressure coefficient. In addition, the pressure coefficient for locally reaching the Prandtl-Glauert limit of $M = 0.3$ is added in the graph. Remembering that a pressure coefficient of $c_p = -3$ relates to a doubled flow velocity in incompressible flow, it is seen that, even for low onflow Mach numbers, a significant influence of compressibility must be expected in high-lift flows.

To underline this, Figure 2.11 shows the evolution of the minimum pressure coefficient at a slat for various subsonic onflow Mach numbers [9]. For the lower Mach numbers $M_\infty = \{0.13; 0.15\}$, the limit minimum pressure is seen for $c_p \approx -25$, indicating a local speed more than five times the onflow speed. For the two higher Mach numbers $M_\infty = \{0.2; 0.25\}$, the evolution of the pressure coefficient differs significantly. The graph indicates the Mayer limit and the critical pressure coefficient for those two Mach numbers. Already at low angles of attack, the pressure coefficient starts to grow faster. Past passing the critical pressure coefficient value, the slope of the curves show a significant kink, and approaching the Mayer limit hinders the pressure coefficient to rise further. It can be concluded that strong compressible effects are present already at an onflow Mach number of $M_\infty = 0.2$, and significant

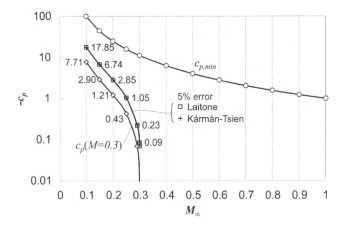

**FIGURE 2.10**   Characteristic values of the pressure coefficient exceeding 5% compressibility correction and where locally the Mach number $M = 0.3$ is reached.

supersonic flow is present at higher angles of attack. For these conditions, the generation of lift is limited by the compressibility of the air. As the onflow Mach number is significantly smaller than the common value obtained from the Prandtl-Glauert correction, it must therefore be recommended to *always* evaluate high-lift airfoil flows in a compressible manner.

## 2.3   THE INFLUENCE OF VISCOSITY

Flow separation is related to flow reversal near a solid wall. This reversal of flow is due to the losses of kinetic energy produced by viscous forces in the boundary layer. This principle was first observed by Ludwig Prandtl [10], the founder of boundary

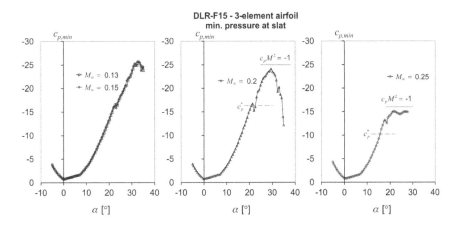

**FIGURE 2.11**   Evolution of minimum pressure coefficient at a slat over angle of attack for various onflow Mach numbers obtained by experiments with the DLR-F15 3-element airfoil [9].

layer theory. He derived from the Navier-Stokes-equations eq. (2.1) the so-called boundary layer equations. Orienting the coordinate system in a way that the x-axis is parallel to the wall and the y-axis is normal to it enables to neglect portions of the Navier-Stokes-equations that are known to be of minor importance. Neglecting all terms based on the main assumptions

$$\rho = const.$$
$$v \ll u$$
$$\frac{\partial v}{\partial x} \ll \frac{\partial u}{\partial y} \tag{2.48}$$
$$\frac{\partial^2 u}{\partial x^2} \ll \frac{\partial^2 u}{\partial y^2}$$
$$\frac{\partial v}{\partial t} \ll 1; \quad \frac{\partial^2 v}{\partial x^2} \ll 1; \quad \frac{\partial^2 v}{\partial y^2} \ll 1$$

in eq. (2.1) leads to the incompressible laminar boundary layer equations [11]

$$\frac{\partial u}{\partial x} + \frac{\partial v}{\partial y} = 0$$
$$\frac{\partial u}{\partial t} + u\frac{\partial u}{\partial x} + v\frac{\partial u}{\partial y} = -\frac{1}{\rho}\frac{\partial p}{\partial x} + v\frac{\partial^2 u}{\partial y^2} \tag{2.49}$$
$$0 = \frac{\partial p}{\partial y}.$$

The major characteristics of the onset of a separation are found at the wall. Due to viscous wall condition, the velocity at the wall is always zero

$$\mathbf{u}\big|_{y=0} = 0. \tag{2.50}$$

Introducing these characteristics into the momentum equation of eq. (2.49) leads to the relation

$$\frac{1}{\rho}\frac{\partial p}{\partial x}\bigg|_{y=0} = \frac{\partial^2 u}{\partial y^2}\bigg|_{y=0}. \tag{2.51}$$

At separation onset, also the velocity above the wall becomes zero and therefore also the velocity gradient perpendicular to the wall

$$\frac{\partial u}{\partial y}\bigg|_{y=0} = 0. \tag{2.52}$$

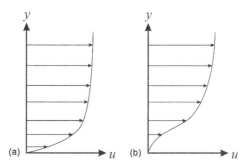

**FIGURE 2.12**  Principal shapes of the velocity distribution in a boundary layer: (a) stable against separation; (b) instable against separation.

This condition leads to the inflection point condition, when a boundary layer gets prone to separation. Figure 2.12 shows the principal shapes of the velocity distribution of the boundary layer.

The left shape is stable in the sense of separation as the velocity gradient off the wall is prominent. In the right figure, the shape is instable as the velocity gradient off the wall tends to zero. Regarding the curvature of the velocity distribution, the characteristic

$$\frac{\partial u}{\partial y} > 0$$

$$\frac{\partial^2 u}{\partial y^2} < 0 \qquad\qquad (2.53)$$

is given in all cases for the outer part of the boundary layer. Therefore, an instable boundary layer is always characterized by an inflection point and a positive curvature off the wall

$$\left.\frac{\partial^2 u}{\partial y^2}\right|_{y=0} > 0 \qquad\qquad (2.54)$$

and due to eq. (2.51), this means that a prerequisite for a flow separation due to viscous effects is a pressure rise along the wall [12].

Figure 2.13 shows the principal properties of a boundary layer flow at separation. Up to the separation point, the wall shear stress is positive as the flow points in the downstream direction close to the wall. At the separation point, the wall shear stress vanishes as the velocity gradient off the wall gets zero. Past the separation, the flow close to the wall is reversed and therefore shows a negative wall shear stress. Hereby, the velocity magnitude is relatively low as the energy for the flow motion is only transferred into the separated flow by viscous forces along with the shear layer. The bounding streamline of the separated flow approximately forms a wedge shape in first-order assumption. The right-hand side shows the development of the velocity distribution normal to the wall in pressure rise. For every streamline, the

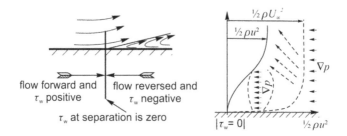

**FIGURE 2.13** The onset of separation due to viscous effects. (Reproduction of Stratford [18], figure 3.)

Bernoulli equation holds and the deceleration of the lower streamlines with reduced flow velocity is stronger than for the outer flow. By this, the development of the instable boundary layer profile with inflection point is obvious.

Some integral values of the velocity distribution of the boundary layer are helpful for its characterization. The displacement thickness $\delta$ is the thickness by which the body must be thickened to obtain the same mass flow in inviscid flow. Figure 2.14 sketches that the area spanned by the velocity at the boundary layer edge $U_e$ and the displacement thickness equals the area of the velocity deficit in the viscous boundary layer.

$$\delta = \frac{1}{U_e} \int_0^\infty (U_e - u)\,dy. \tag{2.55}$$

The momentum loss thickness $\theta$ is analogously the thickness by which a body must be thickened in inviscid flow to compensate for the viscous momentum deficit

$$\theta = \frac{1}{U_e^2} \int_0^\infty u(U_e - u)\,dy. \tag{2.56}$$

A dimensionless number used to characterize the boundary layer is often the Reynolds number in relation to the momentum loss thickness

$$Re_\theta = \frac{U_e \cdot \theta}{\nu}. \tag{2.57}$$

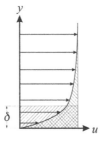

**FIGURE 2.14** Sketch of the determination of the displacement thickness.

The ratio of displacement thickness and momentum loss thickness is called the form factor

$$H_{12} = \frac{\delta}{\theta}.$$ (2.58)

The wall friction coefficient is defined by

$$c_f = \frac{\tau_w}{\frac{\rho}{2}U_e^2} = \frac{2\nu}{U_e^2}\frac{\partial u}{\partial y}\Big|_{y=0}.$$ (2.59)

To obtain an integral view of the velocity profile, the boundary layer equations can be integrated in the wall-normal direction only. By this partial integration, the momentum equation of eq. (2.49) can be transferred into the so-called integral boundary layer equation

$$\frac{d\theta}{dx} + (2 + H_{12})\frac{\theta}{U_e}\frac{dU_e}{dx} = \frac{c_f}{2}.$$ (2.60)

It can be shown that for distributions of the velocity at the boundary layer edge along the wall of the form

$$U_e(x) = kx^m,$$ (2.61)

the shape of the boundary layer along the wall is affine resulting in a constant form factor, which means that a normalization of the form

$$\frac{u}{U_e} = f\left(\frac{y}{\delta}\right)$$ (2.62)

leads to identical velocity profiles. The affinity of the boundary layer velocity profile allows for a coordinate transformation that regards the boundary layer equations in the directions parallel and perpendicular to the wall separately

$$\xi = x; \quad \eta = y\sqrt{\frac{m+1}{\nu}\frac{U_e(x)}{x}}.$$ (2.63)

As in the potential theory, the ansatz of a stream function of the form

$$u = \frac{\partial \Psi}{\partial y}; \quad v = -\frac{\partial \Psi}{\partial x}$$ (2.64)

fulfills the continuum equation. For the boundary layer, the ansatz of the stream function

$$\Psi(x,y) = \sqrt{\frac{\nu}{m+1}xU_e(x)}\,f(\eta)$$ (2.65)

separates the wall-normal distribution from the streamwise variation. The important result of this is the equation of Falkner and Skan [13]

$$2\frac{d^3 f}{d\eta^3} + f\frac{d^2 f}{d\eta^2} + \beta\left(1-\left(\frac{df}{d\eta}\right)^2\right) = 0; \quad \beta = \frac{2m}{m+1},$$ (2.66)

by which the ansatz of the stream function transformed the partial differential equations of eq. (2.49) into an ordinary differential equation. We resume the similarity variable

$$\eta = \frac{y}{\delta_N(x)}$$ (2.67)

with the characteristic boundary layer thickness from the transformation of eq. (2.63) and the ansatz of the velocity distribution along the wall eq. (2.61)

$$\delta_N(x) = \sqrt{\frac{\nu x}{(m+1)kx^m}}.$$ (2.68)

One special solution of the Falkner-Skan equation is the separating boundary layer characterized by

$$\frac{d^2 f}{d\eta^2} = 0.$$ (2.69)

Numerical integration [14] for this condition yields the constants

$$\beta = -0.199; \quad m = -0.091$$ (2.70)

and the corresponding **form factor of the separating laminar boundary layer** profile is

$$H_{12} = 4.029.$$ (2.71)

The low value of $m$ implies that the laminar boundary layer already separates for a relatively low rise of pressure.

As analytically, the velocity distribution eq. (2.61) corresponds to a wedge flow, it was an idealized case for deriving the specific form factor to characterize a separating laminar boundary layer velocity distribution, but it is not predictive for a real airfoil flow with probably turbulent boundary layer. Nevertheless, the approach of normalizing the flow characteristics is suitable here, too. Although airfoil flows do not have in general a velocity distribution that leads to affine boundary layer profiles, an airfoil impacted at the same angle of attack but at different speeds shows affine pressure distributions. A normalization into the so-called canonic pressure coefficient

$$\bar{c}_p = \frac{p - p_\infty}{\underbrace{\frac{1}{2}\rho_\infty U_0^2}_{\text{in incompressible flow}}} = 1 - \left(\frac{U_e}{U_0}\right)^2, \tag{2.72}$$

where $U_0$ is the maximum velocity at the boundary layer edge ahead of the pressure rise, decouples the shape of the pressure distribution from its actual values. The value of the canonical pressure coefficient varies from $\bar{c}_p = 0$ at the start of the pressure rise to $\bar{c}_p = 1$ at a stagnation point. Stratford [15] used this canonical pressure distribution to derive a criterion for flow separation based on the shape of the pressure distribution. Starting for the laminar boundary layer, Stratford separates the velocity profile normal to the wall into two regions. In the outer layer, along a streamline, the shear forces reduce the total pressure of the flow

$$\frac{\partial}{\partial s}\left(p + \frac{1}{2}\rho u^2\right) = \mu \frac{\partial^2 u}{\partial y^2}. \tag{2.73}$$

As the velocity – and therefore the dynamic pressure – is low in the inner layer of a velocity profile close to separation, the shear forces there are mainly counteracting the rise of the static pressure

$$\frac{\partial p}{\partial s} = \mu \frac{\partial^2 u}{\partial y^2}. \tag{2.74}$$

From analysis of the joining conditions and the solution of the Falkner-Skan equation for the velocity profile at separation, Stratford deduces a local **condition for separation of the laminar boundary layer** based on the pressure gradient

$$\bar{c}_p \left(x \frac{\partial \bar{c}_p}{\partial x}\right)^2 = 7.64 \times 10^{-3} \tag{2.75}$$

The advantage of the Stratford criterion over the form factor criterion in eq. (2.71) is that it is a local criterion and doesn't require the full solution of the integral boundary layer equation. The important feature of the Stratford-criterion is that it relates the separation to the pressure gradient of the recovery, not the pressure difference.

    For turbulent boundary layer flows, an analytic relation as the Falkner-Skan equation is not available. Thus, empirical relationships were seeking to establish similar separation requirements. Von Doenhoff and Tetervin [16] derive from experimental data that the **form factor of the turbulent boundary layer at separation is**

$$H_{12} = 1.8 \div 2.3. \tag{2.76}$$

Maskell [17] provides an interactive method to calculate the development of the momentum loss thickness and the form factor from the velocity distribution along the wall. With this, the Ludwieg-Tillmann relation for the wall friction coefficient

$$c_f = 0.246 \cdot e^{-1.561 H_{12}} \cdot Re_\theta^{-0.268} \tag{2.77}$$

is evaluated and the extrapolation of $c_f \to 0$ provides the location of separation.

Stratford performs a similar analysis as for the laminar boundary layer by dividing the velocity profile into an inner and outer layer [18]. From analyzing experimental data [19] Stratford derives a **local criterion when the turbulent boundary layer flow is likely to separate**

$$\frac{\bar{c}_p \left[ x \left( d\bar{c}_p / dx \right) \right]^{1/2}}{\left( 10^{-6} Re_x \right)^{1/10}} = S; \quad S = \begin{cases} 0.39 & \text{if} \quad \dfrac{d^2 p}{dx^2} \ge 0 \ (\text{convex}) \\[2mm] 0.35 & \text{if} \quad \dfrac{d^2 p}{dx^2} < 0 \ (\text{concave}) \end{cases} \quad ; \quad Re_x = \frac{U_0 x}{\nu}. \tag{2.78}$$

It is worth to note that the value of the criterion for the turbulent boundary layer is much higher than for the laminar boundary layer, which reflects the ability of the turbulent boundary layer to better withstand a pressure rise. The criterion is conservative and according to Stratford "the prediction of the pressure rise to separation is likely to be from 0 to 10% too low, which puts it second in accuracy to those methods, such as Maskell's (1951)[5] [...]. However, the convenience of the method makes the present error acceptable for many applications [...]." – (Stratford [18] p. 1.)

An interesting situation appears if the Stratford-criterion is used to derive the pressure distribution that is likely – or about – to separate in the complete pressure rise. Inserting the value of $S = 0.35$ into eq. (2.78) leads to a differential equation. Stratford's solution to this is given by

$$\bar{c}_p(x) = \begin{cases} 0.645 \cdot \left[ 0.435 Re_0^{1/5} \cdot \left[ \left( \frac{x}{x_0} \right)^{1/5} - 1 \right] \right]^{1/3} & \forall \bar{c}_p \le \dfrac{4}{7} \\[4mm] 1 - \dfrac{a}{\left[ \left( \frac{x}{x_0} \right) - b \right]^{1/2}} & \forall \bar{c}_p > \dfrac{4}{7} \end{cases} \tag{2.79}$$

with $x_0$ being the location of the start of the pressure rise and $Re_0$ is the corresponding Reynolds number

$$Re_0 = Re_{x=x_0} = \frac{U_0 x_0}{\nu}. \tag{2.80}$$

The coefficients $a$ and $b$ are evaluated so that the slope of the pressure distribution matches at the joining point, where $\bar{c}_p = \dfrac{4}{7}$.

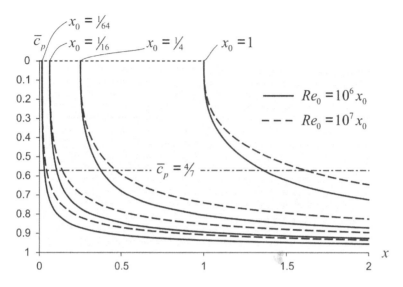

**FIGURE 2.15** Stratford pressure distributions for different locations of pressure rise onset and at different Reynolds numbers. (Reproduction of Smith [5], figure 21.)

Figure 2.15 shows Stratford's canonical pressure distributions for different positions of pressure rise onset $x_0$ and at two different freestream Reynolds numbers. The downstream variation of the pressure rise onset position results in more weakened boundary layers so that the allowable pressure gradient reduces. The same weakening is observed for lower Reynolds numbers as expected. There are some distinct characteristics of these pressure distributions that are to be highlighted with regard to their application to pressure recovery on an airfoil:

- The initial pressure gradient at the pressure rise location tends to infinity $\dfrac{d\overline{c}_p}{dx} \to \infty$ thus revealing that, at the beginning of the pressure rise, no separation can be present.
- The initial shape of the pressure distribution follows approximately $\overline{c}_p \sim (x - x_0)^{1/3}$ thus achieving affine velocity profiles in the beginning (compare to eq. (2.61) for the laminar boundary layer).
- The dominant variable of the pressure distribution is the relative location $x/x_0$ itself, the variation on the onflow Reynolds number is less important.
- Even for the viscous boundary layer, a complete pressure recovery $\overline{c}_p = 1$ is achieved for an infinitely long distance $x/x_0 \to \infty$.
- Following the shape of the Stratford pressure distribution obtains the shortest length of pressure recovery, as at any point the steepest pressure gradient without separation is maintained.
- The Stratford pressure distribution is the one with a minimum drag coefficient. As the boundary condition is the boundary layer being at the limit to separation, this implies that there is approximately no wall shear stress eliminating the contribution of friction to the overall drag.

Robert H. Liebeck (*1937) graduated at the University of Illinois – Urbana Champaign. After graduation, he joined the Douglas Aircraft Company where he worked on his PhD under supervision of A.M.O. Smith.

One of his major contributions to aerodynamics has been the development of the methodology to design airfoils obtaining maximum lift outlined in his PhD thesis. He used the Stratford pressure distribution for the inverse design of airfoils as the property of shortest pressure recovery without separation allows for keeping the suction pressure level high until the most downstream position. The so-called Liebeck-airfoils are even today in use for wind energy, ventilation fans, and racing car applications.

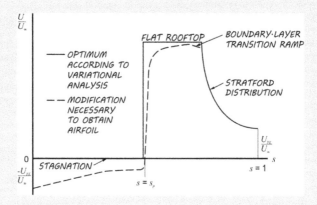

Pressure distribution for maximum lift coefficient airfoil by inverse design.

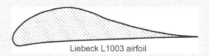

Typical shape of a Liebeck airfoil.

**Sources**: Aero-Astro, no. 4, MIT, 2007; Liebeck RH (1978) Design of Subsonic Airfoils for High Lift, AIAA Journal of Aircraft **15**(9), pp. 547–561.

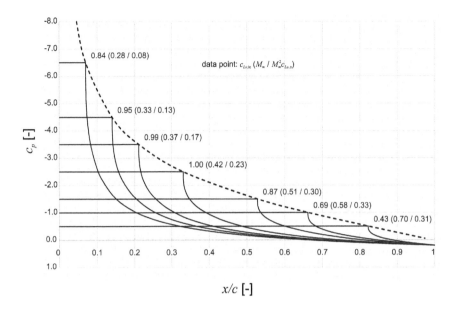

**FIGURE 2.16** Turbulent roof-top pressure distributions with indicated contribution of the upper side to the lift coefficient for $Re_c = 5 \times 10^6$. (Reproduction from Smith [5], figure 25.)

The property of the shortest length for pressure recovery makes this pressure distribution of the recovery highly interesting to obtain airfoils that provide a maximum lift coefficient. It allows shifting the start of the pressure recovery as far downstream as possible. As an example, for a pure subsonic airfoil, the local pressure coefficient has to be less than the critical pressure coefficient $c_p^*$. The resulting shape of a pressure distribution to achieve a maximum lift coefficient is ideally a roof-top profile with a Stratford recovery [20].

Figure 2.16 shows such turbulent roof-top pressure distributions for the upper side of an airfoil with a pressure recovery to $c_{p,TE} = 0.2$ for different levels of the suction pressure coefficient at the beginning of the pressure rise. The indicated numbers are the contribution of the upper side to the section lift coefficient

$$c_{L_u} = -\int_0^1 c_p dx. \tag{2.81}$$

The maximum lift coefficient is obtained for a pressure level of $c_{p,\min} \approx -2.5$.

When it comes to lift force instead of lift coefficient, it is necessary to respect the limit on the onflow velocity, since the lift force scales with both.

$$L \sim M_\infty^2 c_{L_u}, \tag{2.82}$$

Assuming that the flow shall stay subsonic, the Mach number of the onflow is related to the suction pressure level so that the pressure coefficient doesn't exceed

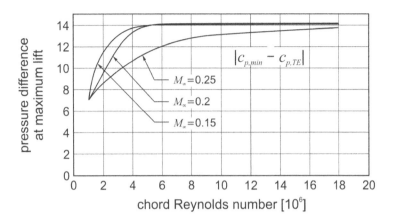

**FIGURE 2.17** Pressure differences between suction peak and trailing edge obtained for different Mach and Reynolds numbers from experimental data of the 30P30N airfoil. (Reproduction from Valarezo & Chin [21], figure 2.)

the critical pressure coefficient. It is worth to note that the compressibility correction according to eq. (2.46) or (2.47) needs to be applied. In Figure 2.16, the numbers in brackets list the onflow Mach number and the corresponding contribution to lift calculated applying the Kármán-Tsien compressibility correction[6]. It can be seen that the higher lift force is generated at higher Mach numbers than the maximum lift coefficient itself.

Another type of criterion to address the limitation of lift generation found in the literature is the semi-empirical pressure difference criterion by Valarezo and Chin [21]. They analyzed numerous experimental data obtained with the 30P30N high-lift airfoil in various configurations and at different Mach and Reynolds numbers. They obtained a correlation of the pressure difference of the pressure rise and derived the relation known as the pressure difference rule. Figure 2.17 shows the summary graph of Valarezo and Chin, where the maximum pressure difference measured at the 30P30N airfoil for different Mach numbers is plotted versus Reynolds number. Their analysis shows a clear trend for a saturation of the maximum pressure difference for high Reynolds numbers at

$$\Delta c_p = |c_{p,\min} - c_{p,TE}| < 14. \qquad (2.83)$$

Hereby, at higher Mach numbers the saturation is at higher Reynolds numbers. For Mach numbers less than $M_\infty \leq 0.2$, the maximum pressure difference is achieved at Reynolds numbers at about $Re_\infty \approx 5 \times 10^6$.

Nevertheless, measurements on another high-lift airfoil, the L1T2 airfoil [22], revealed different values, namely

$$\Delta c_p < 11 \quad \text{(wing)}$$
$$\Delta c_p < 22 \quad \text{(slat)} \qquad (2.84)$$

and even more, recent measurements on the DLR-F15 airfoil [9] obtained

$$\Delta c_p < 14 \div 16 \quad (\text{wing})$$
$$\Delta c_p < 19 \div 22 \quad (\text{slat})$$

(2.85)

depending on the chord of the wing or slat. Concluding, it has to be stated that the viability of a pressure difference criterion for the estimation of maximum lift is limited. It may be applicable for multi-element airfoils when only the positioning of the elements is changed, but there is a severe influence of the basic shape and element sizes that prevents the universality of this type of criterion.

## 2.4   TYPES OF WING STALL

For a pilot, the normal flight aircraft reaction on a pitch up is the reduction of the airspeed. In the condition when the lift coefficient is no longer increasing with increasing angle of attack, the flight speed increases again, as the lift force gets lower than the aircraft weight. The aircraft stalls.

Classically, the type of stall was classified into categories depending on leading-edge radius and Reynolds number. The major presumption of the explanation of the occurrence of the different types is based on a laminar boundary layer up to the suction peak at the leading edge, a following laminar separation bubble where the flow changes its state into turbulent and reattaches after a short length. Woodward et al. [28] summarize the different stall types. Figure 2.18 reproduces their graph with the stall pattern seen by the lift curve and Figure 2.19 shows the corresponding pressure distributions and flow patterns. The three initial types of airfoil stall introduced by McCullogh and Gault [23] were:

1. *"Trailing-edge stall (preceded by movement of the turbulent separation point forward from the trailing edge with increasing angle of attack)"*

   **(McCullough & Gault [23] p. 4)**

   Trailing-edge stall is the most common for airfoils with rounded leading edge at sufficient high Reynolds numbers. Due to the friction losses of energy in the turbulent boundary layer, the flow is incapable to run against the pressure rise towards the trailing edge and separation occurs (point A4 in Figure 2.18/Figure 2.19(a)). "With increasing angle of attack the separation point moves upstream. The peak of the lift curve is rounded and the loss of lift as well as the increase of pressure drag after the stall is gradual" – (McCullough & Gault [23] p. 5). For the investigated airfoil, McCullogh and Gault concluded that "at maximum lift the flow was separated over approximately the rear half of the airfoil (point B4 in Figure 2.18/Figure 2.19(a)). Beyond maximum lift the forward progression of separation continued at about the same rate as prior to the stall" – (McCullough & Gault [23] p. 5).

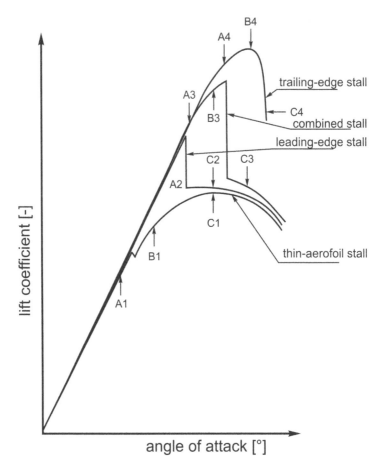

**FIGURE 2.18** Typical lift curves for four stall patterns. (Reproduced from Woodward et al. [28], figure 1.)

2. *"Leading edge stall (abrupt flow separation near the leading edge generally without subsequent reattachment)"*

**(McCullough & Gault [23] p. 4)**

Leading edge stall occurs for thin airfoils with rounded leading edge at moderate Reynolds numbers ($1 \times 10^6 < Re_c < 1 \times 10^7$). The lift characteristics show a sudden lift breakdown after reaching maximum lift (point C2 in Figure 2.18/Figure 2.19(b)) with no prior indication of stall onset. The stall mechanism consists of a sudden burst of the laminar separation bubble when the flow is not able to reattach after transition to turbulent flow. A further characteristic of the laminar bubble burst is "an abrupt collapse of the leading-edge pressure peak" – (McCullough & Gault [23] p. 7) – leading to a flat suction plateau of the constant pressure region of the separated flow with only a small leading suction pressure peak.

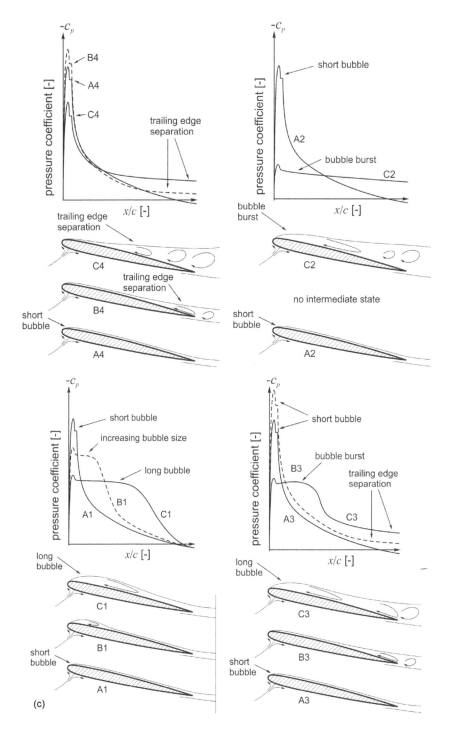

**FIGURE 2.19** Pressure distributions and boundary layer states through stall. (a) Trailing edge stall; (b) leading-edge stall; (c) thin-airfoil stall; (d) combined stall (Reproduced from Woodward et al. [28], figure 2(a)–(d).)

3. *"Thin-airfoil stall (preceded by flow separation at the leading edge with reattachment at a point which moves progressively rearwards with increasing angle of attack)"*

**(McCullough & Gault [23] p. 5)**

Thin-airfoil stall occurs for sharp-edged airfoils or rounded leading edges at very low Reynolds numbers. The laminar separation bubble extends and the reattachment point moves towards the trailing edge (point B1 in Figure 2.18/Figure 2.19(c)). In the lift curve, this is characterized by a smooth reduction of the lift gradient $\partial C_L/\partial\alpha$. Stall occurs "when the reattachment point coincides with the trailing edge" – (McCullough & Gault [23] p. 11; point C1 in Figure 2.18).

Later, Gault [24] introduced a fourth stall type as the combined stall – the combination of leading edge and trailing-edge stall. It is characterized by a starting trailing-edge separation – visible by the reduction of the lift gradient (point B3 in Figure 2.18/Figure 2.19(d)) – and a leading edge separation (point C3 in Figure 2.18) before the trailing-edge separation fully evolves to airfoil stall.

Gault classified the occurrence of the four different stall types based on the Reynolds number and a geometric parameter of the leading edge bluntness, the upper surface ordinate at $x/c = 1.25\%$. Although "the relationship between stalling characteristics and an upper surface ordinate near the leading edges of airfoils has no apparent physical significance [,] the degree of correlation obtained [...] is surprising" – (Gault [24] p. 5). Figure 2.20 resembles the appearance of the different

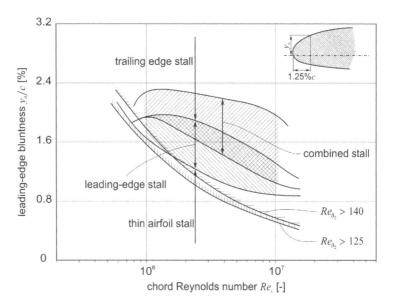

**FIGURE 2.20** Types of stall in dependence on leading edge bluntness and Reynolds number by Gault [24]. (Reproduced from Van den Berg [25], figure 3.)

stall types depending on the Reynolds number and the leading edge bluntness. For low Reynolds numbers, less than $Re_c < 1 \times 10^6$, thin-airfoil stall dominates even for thicker or blunter leading edges. For a Reynolds number higher than $Re_c > 1 \times 10^7$, mostly trailing-edge stall occurs on a single airfoil.

There was a long-time debate on the existence of a turbulent leading edge separation. According to the integral boundary layer solution by Thwaites [26], analysis of the integral boundary layer equation by Van den Berg [25] concluded that laminar separation occurs for

$$\lambda = \frac{\delta_2^2}{\nu} \frac{\partial U_e}{\partial s} = -0.09 \qquad (2.86)$$

and laminar bubble burst should only occur for a momentum thickness Reynolds number at the separation point beyond $Re_{\delta_2} < 125^7$ or $Re_{\delta_2} < 140^8$. These boundaries are introduced in Figure 2.20 and show only a small region of the leading edge separation that falls into the region of sufficiently low Reynolds numbers. Van den Berg further showed pressure distributions of an abrupt stall at approximately $x/c = 10 \div 15\%$ without the full pressure plateau of the laminar leading edge separation. He introduced the non-dimensional separation parameter

$$a^+ = \frac{\rho^{1/2} \nu}{\tau_w^{3/2}} \frac{dp}{ds} \qquad (2.87)$$

as the ratio of the pressure forces to the shear forces in-wall quantities. The analysis from measurements of two airfoils at high angles of attack suggests that turbulent leading edge separation occurs at $a^+ > 1.5 \div 2 \times 10^{-2}$.

A last stall type is not related to viscous effects but to compressibility. It especially appears at airfoils or wings with flaps. Figure 2.21 (a) shows corresponding lift and drag curves of a 2-element airfoil at moderate onflow Mach number $M_\infty = 0.18$ at flight Reynolds number $Re_c = 22.8 \times 10^6$. The lift characteristic shows an unusually smooth stall over a large range of incidences. The drag characteristics show an increase more in kind of a step than the strong gradual drag increase associated with flow separation onset. Detailed analyses by numerical simulation unveiled that no separation was visible for this airfoil until the highest angle of attack $\alpha < 12°$, where the stall got more abrupt. The origin of the stall behavior was then detected by a very close look to the leading edge.

Figure 2.21 (b) shows the local Mach number at the leading edge of the main airfoil at the highest angle of attack without flow separation of $\alpha = 11°$. The flow field exhibits a significant supersonic flow region just at the leading edge with a local Mach number up to $M = 1.3$. The supersonic region ends with a straight shock, which is responsible for the limit in achievable lift coefficient. According to transonic flow theory, the flow speed past the straight shock is always subsonic. The faster the flow is in the supersonic region ahead of the shock, the lower the flow speed past the shock, according to the shock relations [1]. The larger the acceleration around the leading edge, the stronger the shock and the lower the flow speed along the upper surface. In conclusion, the pressure level on the upper side rises and leads

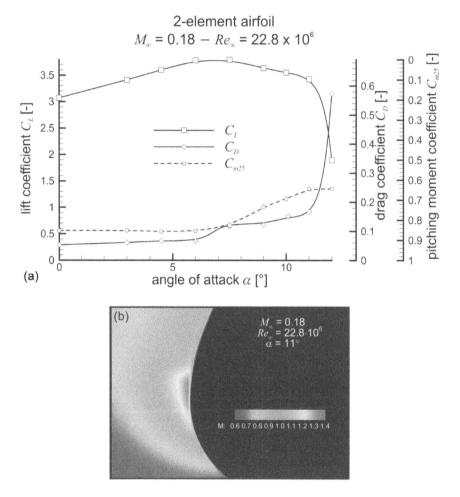

**FIGURE 2.21** Airfoil stall due to compressibility of a 2-element airfoil: (a) lift and drag characteristics over angle of attack; (b) flow field visualization of local Mach number contours at the wing leading edge post stall.

to lift reduction. It is important to point out a major difference to the flow around a transonic airfoil in cruise: the orientation of the shock is not orthogonal to the flow direction along the upper airfoil surface but parallel to the airfoil chord. This makes the shock nearly invisible in the pressure distribution if plotted as common along the non-dimensionless chord axis $x/c$. Additionally, the compression losses, which are usually known as wave drag, act vertically and, therefore, not in the drag direction. The only contribution seen in the drag coefficient is therefore the increased pressure drag by thickening the boundary layer due to shock-boundary layer interaction.

## NOTES

1. From the huge source of literature, the following is the descriptions as given by Schlichting & Truckenbrodt [1], Anderson [2], and Radespiel [3].
2. The notation of the potential theory is following the notation by Katz & Plotkin [4].
3. It is important to highlight that this is the Kutta-condition at whole. It is derived from the observation that an attached flow does not go around the trailing edge. Popular interpretations as the flow off the airfoil tangential to the trailing edge or the equilibrium of upper and lower side pressure are only consequences of this condition.
4. The original report from Mayer [6] is still non-disclosed as NACA-Limited to US Government and Contractor use only per 2013 NTRS Review, as identified by the Listing of National Advisory Committee for Aeronautics (NACA) technical reports designated or identified in the National Aeronautics and Space Administration (NASA) Scientific and Technical Information (STI) database as restricted or nonpublic, 1944–1959 (2014).
5. The noted reference guides to an unpublished paper of Maskell, probably a pre-release of [17].
6. Smith [5] used the Kármán-Tsien rule to calculate the correct onflow Mach number, but the pressure distribution itself was not scaled with the compressibility. Thus, the values of the reproduced graphics differ from the original.
7. Following Thwaites' integral boundary layer solution [26].
8. Following Gaster's integral boundary layer solution [27].

## REFERENCES

[1] Schlichting H, Truckenbrodt E (2001), Aerodynamik des Flugzeuges, Band 1. Springer Verlag, Berlin, Heidelberg, New York.
[2] Anderson JD Jr. (1991) Fundamentals of Aerodynamics, 2nd edition, McGraw Hill, New York.
[3] Radespiel R (2007) "Manuscript of lecture 'Airfoil Aerodynamics'," TU Braunschweig.
[4] Katz J, Plotkin A (2001) Low Speed Aerodynamics, 2nd edition, Cambridge University Press, Cambridge, UK.
[5] Smith AMO (1975) "High-Lift Aerodynamics," Journal of Aircraft 12(6), pp. 501–530.
[6] Mayer JP (1948) "A Limit Pressure Coefficient and an Estimation of Limit Forces on Airfoils at Supersonic Speeds," NACA Research Memo L8F23.
[7] Von Karman T (1941) Compressibility Effects in Aerodynamics, Journal of the Aeronautical Sciences 8(9), pp. 337–356, DOI: 10.2514/8.10737.
[8] Laitone EV (1951) New Compressibility Correction for Two-Dimensional Subsonic Flow, Journal of the Aeronautical Sciences 18(5), p. 350, DOI:10.2514/8.1951.
[9] Wild J (2013) Mach and Reynolds Number Dependencies of the Stall Behavior of High-Lift Wing-Sections, Journal of Aircraft 50(4), pp. 1202–1216.
[10] Prandtl L (1904) "Über die Flüssigkeitsbewegung bei sehr kleiner Reibung," III International Mathematiker-Kongress, Heidelberg.
[11] Schlichting H, Gersten K (2006) Grenzschichttheorie, 10th edition, Springer Verlag, Berlin Heidelberg.
[12] Radespiel R (2007) "Skript zur Vorlesung Strömungsmechanik 2," TU Braunschweig.
[13] Falkner VM, Skan SW (1930) Some Approximate Solutions of the Boundary Layer Equations. Phil. Mag. 12(1931), pp. 805–896; ARC-Report 1314.
[14] Hartree DR (1937) On an Equation Occurring in Falkner and Skan's Approximate Treatment of the Equations of the Boundary Layer, Proceedings of the Cambridge Philosophical Society, 33(2), pp. 223–239.

[15] Stratford BS (1957) Flow in the Laminar Boundary Layer near Separation, ARC R&M 3002.
[16] Von Doenhoff AE, Tetervin N (1943) Determination of General Relations for the Behavior of Turbulent Boundary Layers, NACA TR 772.
[17] Maskel EC (1951) Approximate Calculation of the Turbulent Boundary Layer in Two-Dimensional Incompressible Flow, RAe report AERO.2334.
[18] Stratford BS (1959) The Prediction of Separation of the Turbulent Boundary Layer, Journal of Fluid Mechanics 5(1), pp. 1–16.
[19] Stratford BS (1959) An Experimental Flow with Zero Skin Friction throughout its Region of Pressure Rise, Journal of Fluid Mechanics 5(1), pp. 17–35.
[20] Liebeck RH (1978) Design of Subsonic Airfoils for High Lift, AIAA Journal of Aircraft 15(9), pp. 547–561.
[21] Valarezo WO, Chin VD (1994) Method for the prediction of Wing Maximum Lift, Journal of Aircraft 31(1), pp. 103–109.
[22] Woodward DS, Lean DE (1993) Where Is High-Lift Today? – A Review of Past UK Research Programmes, no. 1 in High-Lift System Aerodynamics, AGARD CP 515.
[23] McCullogh GB, Gault DE (1951) Examples of Three Representative Types of Airfoil-Section Stall at Low Speed, NACA TN 2502.
[24] Gault DE (1957) A Correlation of Low-Speed, Airfoil-Section Stalling Characteristics with Reynolds Number and Airfoil Geometry, NACA TN 3963.
[25] Van den Berg B (1981) Role of Laminar Separation Bubbles in Airfoil Leading edge Stalls, Journal of Aircraft 19(5), pp. 553–556.
[26] Thwaites B (1949) Approximate Calculation of the Laminar Boundary Layer, The Aeronautical Quarterly 1(3), pp. 245–280.
[27] Gaster M (1969) The Structure and Behavior of Laminar Separation Bubbles. ARC R&M 3595.
[28] Woodward DS, Hardy BC, Ashill PR (1988) Some Types of Scale Effect in Low-Speed, High-Lift Flows, 16th Congress of the International Council of the Aeronautical Sciences, ICAS-88-4.9.

# 3 Airworthiness

## NOMENCLATURE

| | | | | | |
|---|---|---|---|---|---|
| $c$ | $m$ | Reference chord length | $V_{MC}$ | $m/s$ | Minimum control speed |
| $F_z$ | $N$ | Centrifugal force | $V_{MCG}$ | $m/s$ | Minimum control speed on ground |
| $g$ | $m/s^2$ | Gravity acceleration | $V_{MCL}$ | $m/s$ | Minimum control speed in landing configuration |
| $L$ | $N$ | Lift force | $V_{MCL-2}$ | $m/s$ | Minimum control speed in landing configuration with second engine failure |
| $n_z$ | – | Load factor | $V_{MU}$ | $m/s$ | Minimum unstick speed |
| $n_{zw}$ | – | Load factor normal to flight path | $V_R$ | $m/s$ | Rotation speed |
| $q$ | $Pa$ | Dynamic pressure | $V_{REF}$ | $m/s$ | Reference approach speed |
| $R$ | $m$ | Flight path radius | $V_S$ | $m/s$ | Stall speed |
| $s$ | $m$ | Gust penetration length | $V_{SR}$ | $m/s$ | Reference stall speed |
| $S$ | $m^2$ | Wing area | $V_{SW}$ | $m/s$ | Stall warning speed |
| $U_{ds}$ | $m/s$ | Design velocity | $V_{S1g}$ | $m/s$ | 1-g stall speed |
| $U_{gust}$ | $m/s$ | Gust velocity | $V_1$ | $m/s$ | Decision speed |
| $V$ | $m/s$ | Flight speed | $V_2$ | $m/s$ | Climb speed |
| $V_{CLMAX}$ | $m/s$ | Maximum lift coefficient speed | $V_{2min}$ | $m/s$ | Minimum climb speed |
| $V_{EF}$ | $m/s$ | Engine failure speed | $w$ | $m/s$ | Climb/sink rate |
| $V_F$ | $m/s$ | Flap design speed | $W$ | $m/s$ | Aircraft weight |
| $V_{FE}$ | $m/s$ | Flap extended speed | $\gamma$ | $m/s$ | Flight path angle |
| $V_{FTO}$ | $m/s$ | Final take-off speed | $\phi$ | $m/s$ | Bank angle |
| $V_{LOF}$ | $m/s$ | Lift-off speed | | | |

An aircraft is a complex technical system with – in case of failure – a high impact potential. Aircraft manufacturers must, therefore, demonstrate that their product is able to fly safely and that all measures are taken to minimize the risk of failure. As take-off and landing are very critical phases of flight, safety concerns directly affect the requirements set to the high-lift system. In this chapter, we will see how regulations asking for making flying a safe transport affect the design requirements to the high-lift system.

Aircraft have to be certified to fulfill all requirements regarding the properties and handling with respect to the regulations issued by the respective authorities. This certificate concludes the *airworthiness of an aircraft type*. During the life cycle of the aircraft, the obligation to keep the aircraft in a condition that the airworthiness is not affected is moved to the owner. The owner has to inspect and prove the *continuing airworthiness of the aircraft* by documentation and audits.

DOI: 10.1201/9781003220459-3

Air transportation is a global business. To be allowed to fly from country A to country B, it is necessary to fulfill the airworthiness requirements of both countries and all additional countries that may be overflown. Fortunately, agreements are in place, which harmonizes the airworthiness requirements of most countries. Most countries accept the airworthiness in case of a proven certification by the United States and/or Europe on a bilateral basis.

In Europe, the relevant authority is the European Aviation Safety Agency (EASA) located in Cologne, Germany. It issues the so-called Certification Specifications (CS) for all issues related to air transportation. Before the foundation of the EASA, these regulations were known as Joint Aviation Regulations (JAR) and were in place until 2003. In the United States, the relevant authority is the Federal Aviation Administration (FAA). Their regulations are introduced in the Code of Federal Regulations (CFR) as Title 14 "Aeronautics and Space," also known as Federal Aviation Regulations (FAR). Although both regulations are not identical, they are structured in a similar manner that simplifies the search for specific regulations. Nevertheless, the contents of the regulations can differ in detail.

The regulations are subdivided into parts. The airworthiness standards contained therein are associated to different parts with respect to the aircraft category. Table 3.1 lists the defined aircraft categories and the respective parts of the FAA and EASA regulations. The given criteria classify the applicable certification rules. To fall into a certain category, all criteria must be fulfilled, otherwise the aircraft falls into the next higher category.

Very interesting is that for gliders, motor gliders, and very light aircrafts, the FAR regulations condense their airworthiness into one single paragraph: "For special classes of aircraft, including the engines and propellers installed thereon (e.g. gliders, airships, and other nonconventional aircraft), for which airworthiness standards have not been issued under this subchapter, the applicable requirements will be portions of those airworthiness requirements contained in parts 23, 25, 27, 29, 31, 33, and 35 found by the FAA to be appropriate for the aircraft and applicable to a

## TABLE 3.1
## Aircraft Categories and Corresponding Airworthiness Regulations

| Category | Max. Aircraft Mass | No. of Passengers[a] | Propulsion | CFR 14 | EASA |
|---|---|---|---|---|---|
| Glider | ≤750 kg | ≤1 | None | Part 21 | CS-22 |
| Motor glider | ≤850 kg | ≤1 | 1 Piston engine | (21.17(b)) | |
| Very light aircraft | ≤750 kg | ≤1 | 1 Piston engine | | CS-VLA |
| Normal, aerobatic, and utility aircraft | ≤5670 kg (12500 lb) | ≤9 | n.a. | Part 23 | CS-23 |
| Commuter aircraft | ≤8618 kg (19000 lb) | ≤19 | ≥2 Propeller | | |
| Large transport aircraft | n.a. | n.a. | n.a. | Part 25 | CS-25 |

[a] Without crew

specific type design, or such criteria as the FAA may find provide an equivalent level of safety of those parts." – (CFR 14 [1] Part 21.0017(b)). This means that, for such aircrafts, there is no distinct FAA certification procedure, but the rules to follow for the specific aircraft are to be negotiated with the authority.

In the following, we concentrate on the certification rules for large transport aircraft according to Part 25. Throughout this chapter, the actual versions of the airworthiness regulations at the time of writing are cited [1]. Naturally, the regulations can be subject to change anytime. It is therefore required to look for the actual version during an aircraft design. For aircrafts with an existing certification, always the version is relevant at the time of certification. It is not enforced to repeat a certification procedure and to show compliance with changing regulations afterward.

## 3.1 THE DEFINITION OF THE STALL SPEED

Related to high-lift systems, the determination of the stall speed is the most relevant paragraph. The complete definition of flight conditions in any paragraph concerning high-lift systems is done via over-speed ratios related to the stall speed. The definition of the stall speed is regulated by paragraph §25.0103 (a), nearly identically according to CS-25 and FAR 25:

§25.0103 **Stall speed**

a. The reference stall speed $V_{SR}$ is a calibrated airspeed defined by the applicant. $V_{SR}$ may not be less than a 1-g stall speed. $V_{SR}$ is expressed as:

$$V_{SR} \geq \frac{V_{CLMAX}}{\sqrt{n_{zw}}}$$

where –

$V_{CLMAX}$ = Calibrated airspeed obtained when the load factor-corrected lift coefficient $\left(\frac{n_{zw}W}{qS}\right)$ is first a maximum during the manoeuvre [FAR 25: maneuver] prescribed in subparagraph [FAR 25: paragraph] (c) of this paragraph [FAR 25: section]. In addition, when the manoeuvre [FAR 25: maneuver] is limited by a device that abruptly pushes the nose down at a selected angle of attack (e.g., a stick pusher), $V_{CLMAX}$ may not be less than the speed existing at the instant the device operates;

$n_{zw}$ = Load factor normal to the flight path at $V_{CLMAX}$;
$W$ = Aeroplane gross weight;
$S$ = Aerodynamic reference wing area; and
$q$ = Dynamic pressure.

**(CS-25, past Amendment 18 [2], FAR 25 past Amendment 25–108 [3])**

The so-called reference stall speed that all other paragraphs reference is therefore **selected** by the manufacturer. The reference stall speed has to be demonstrated during flight tests by the aircraft manufacturer to be larger or equal to the load factor-corrected airspeed where the aircraft stalls or a stall prevention mechanism actively prevents the aircraft stalling.

It is highlighted here that the definition of the stall speed changed in the year 2002 [3]. Formerly, the paragraph §25.0103 (a) read [4]: "$V_S$ is the calibrated stalling speed, or the minimum steady flight speed, in knots, at which the airplane is controllable." – (CS-25, Amendment 13 [5], FAR 25 Amendment Doc. No. 5066 [4]) – and had to be demonstrated in a certain maneuver. Since the dynamics of the maneuver was not taken into account, the stall speed was therefore always slightly lower than with the new definition. To account for this, all speed definitions that relate to the stall speed by over-speed ratios were adopted in the same amendment by reducing the safety factors.

## 3.2  SPEED DEFINITIONS

Table 3.2 lists the speed definitions relevant for high-lift systems that are applicable for every high-lift configuration, independent of take-off, approach, or landing flight. The stall warning speed, $V_{SW}$, has to be defined by the manufacturer and is the speed when the stall warning has to be activated when approaching stall condition. The stall warning speed has to be at least 5% higher than the stall speed but at least 2 kts calibrated airspeed (CAS).

The upper end of flight speeds with deflected high-lift system is given by the flap design speed, $V_F$, and the flap extended speed, $V_{FE}$. The flap design speed "must be sufficiently greater than the operating speed recommended for the corresponding stage of flight" – (§25.0335(e)(1)). $V_F$ may not be less than 1.6 times the 1-g stall speed, $V_{S1g}$, "with the wing-flaps in take-off position at maximum take-off weight" – (§25.0335(e)(3)(i)), and 1.8 times the 1-g stall speed, $V_{S1g}$, "with the wing-flaps in approach position" – (§25.0335(e)(3)(ii)) – or "in landing position" – (§25.0335(e)(3)(iii)). The flap extended speed, $V_{FE}$, "must be established so that it does not exceed the design flap speed," $V_F$ – (§25.1511).

### 3.2.1  TAKE-OFF AND CLIMB

The speeds listed in Table 3.3 are defined for the take-off flight phase. The sequence of how these speeds are achieved during a normal take-off is depicted in Figure 3.1. Some speeds are not directly visible in the take-off procedure and therefore omitted.

---

**TABLE 3.2**

**General Speed Definitions for High-Lift Systems according to CS-25/FAR 25 and Corresponding Paragraphs**

| Name | Description | Paragraph |
|------|-------------|-----------|
| $V_{SW}$ | Stall warning speed | §25.0203(c) |
| $V_F$ | Flap design speed | §25.0335 |
| $V_{FE}$ | Flap extended speed | §25.1511 |

---

**TABLE 3.3**

**Speed Definitions for Take-Off according to CS-25/FAR 25 and Corresponding Paragraphs**

| Name | Description | Paragraph |
|------|-------------|-----------|
| $V_{MC}$ | Minimum control speed | §25.0149 |
| $V_{MCG}$ | Minimum control speed on ground | §25.0149(e) |
| $V_{MU}$ | Minimum unstick speed | §25.0107(d) |
| $V_{EF}$ | Engine failure speed | §25.0107(a)(1) |
| $V_1$ | Decision speed | §25.0107(a)(2) |
| $V_R$ | Rotation speed | §25.0107(e) |
| $V_{LOF}$ | Lift-off speed | §25.0107(f) |
| $V_2$ | Climb speed | §25.0107(c) |
| $V_{2\,min}$ | Minimum climb speed | §25.0107(b) |
| $V_{FTO}$ | Final take-off speed | §25.0107(g) |

The defined speeds have to be certified for every high-lift configuration setting usable during take-off, and additionally for icing conditions.

The minimum control speed, $V_{MC}$, "is the calibrated airspeed, at which, when the critical engine is suddenly made inoperative, it is possible to maintain control of the airplane with that engine still inoperative, and maintain straight flight with an angle of bank of not more than 5°" – (§25.0149(b)). Therefore, this is the speed when the forces on all control surfaces allow for a safe maneuvering of the aircraft even in the case of engine failure. This speed has to be demonstrated by flight tests. An additional requirement is that $V_{MC}$ may not exceed the reference stall speed by more than 13% – (§25.0149(c)). The selection of the minimum control speed sizes the control surfaces. Nevertheless, it is wise to select a minimum control speed below the stall speed to maintain full control over the aircraft even in cases approaching stall condition with an engine inoperative.

The minimum control speed on ground, $V_{MCG}$, is similar but only affecting the directional control. It "is the calibrated airspeed during the take-off run at which, when the critical engine is suddenly made inoperative, it is possible to maintain control of the aeroplane using the rudder control alone (without the use of nosewheel steering), [...], and the lateral control to the extent of keeping the wings level [...]" – (§25.0149(d)). It is the minimum speed when the nose wheel is no longer needed

**FIGURE 3.1**   Take-off path and related speeds.

to keep the aircraft on the runway. It mainly sizes the rudder control surfaces and the fin.

The minimum unstick speed, $V_{MU}$, "is the calibrated airspeed at and above which the aeroplane can safely lift off the ground, and continue the take-off" – (§25.0107(d)). It is at least the higher of $V_{MC}$ and $V_{SR}$. This speed reflects that the aircraft may be limited in angle of attack. For very long fuselages or low landing gear lengths, the aircraft may be limited when the tail hits the ground (tail strike). In such cases, it may occur that the reference stall speed is not achievable on ground as it may require a higher angle of attack. Similarly, if the efficiency of control surfaces used to demonstrate the minimum control speed requires higher incidences, the minimum unstick speed may have to be increased.

The lift-off speed, $V_{LOF}$, "is the calibrated airspeed at which the aeroplane first becomes airborne" – (§25.0107(f)). The lift-off speed must at least be 10% higher than "$V_{MU}$ in the all-engines-operating condition," and 5% higher than "$V_{MU}$ determined at the thrust-to-weight ratio corresponding to the one-engine-inoperative condition" – (§25.0107(e)(1)(iv)(A)). In case the $V_{MU}$ is limited by the geometry (tail strike), the speed reserve factors are reduced to 8% in the all-engines-operating case and 4% in the one-engine-inoperative condition – (§25.0107(e)(1)(iv)(B)).

The engine failure speed, $V_{EF}$, and the decision speed, $V_1$, are related to safety in the occurrence of an engine failure, while the aircraft is still accelerating on ground. Both speeds are defined by the aircraft manufacturer. They define the needed runway length. The decision speed has to be demonstrated that once the aircraft has achieved this speed, the thrust of the remaining engine(s) is still sufficient to enable a safe take-off and climb. The engine failure speed has to be chosen accordingly by the manufacturer as the velocity, at which the engine may fail in so-called accelerate-stop tests and "may not be less than $V_{MCG}$" – (§25.0107(a)(1)) – and it should be less than a speed at which – without action of the pilot – the aircraft may reach after engine failure $V_1$.

The rotation speed, $V_R$, is the speed at which the aircraft can safely lift the nose landing gear off the ground – (§25.0111(b)). The rotation speed may not be less than the decision speed $V_1$ – (§25.0107(e)(1)(i)), as it should already be safe to continue the take-off. It must be 5% higher than $V_{MC}$, as it should be safe to only rely on the aerodynamic control surfaces to fully control the aircraft – (§25.0107(e)(1)(ii)). Further on, the speed must be selected so that, even with the highest rotation rate, the lift-off does not occur below the minimum lift-off speed, $V_{LOF}$ – (§25.0107(e)(1)(iii)).

The climb speed, $V_2$, is the speed that needs to be achieved at an altitude of 35 ft guaranteeing a sufficient climb rate – (§25.0107(c)). The over-speed ratio of the minimum climb speed, $V_{2min}$, regarding the reference stall speed is depending on the propulsion type. For two and three engine turbo-propeller powered aircraft and turbojet aircraft that have no "provisions for obtaining a significant reduction in the one-engine-inoperative power-on stall speed," the over-speed needs to be 13% above $V_{SR}$ – (§25.0107(b)(1)). For four engine turbo-propeller powered aircraft and turbojet aircraft that have such "provisions for obtaining a significant reduction in the one-engine-inoperative power-on stall speed," the needed over-speed ratio is reduced to 8% above $V_{SR}$ – (§25.0107(b)(2)). Additionally, $V_{2min}$ has to be at least 10% higher than $V_{MC}$ – (§25.0107(b)(3)).

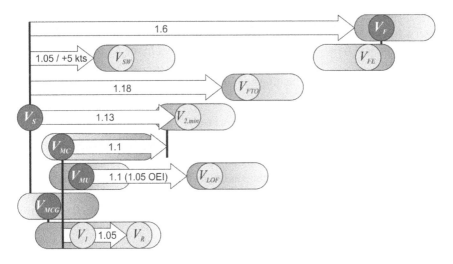

**FIGURE 3.2**   Relation of speed definitions during take-off.

The final take-off speed, $V_{FTO}$, is the speed at which the configuration can be transitioned to the en-route configuration, and a still sufficient climb gradient can be achieved. $V_{FTO}$ must at least be 18% higher than $V_{SR}$ – (§25.0107(g)(1)).

The relations between the defined speeds during take-off are graphically summarized in Figure 3.2.

### 3.2.2   Approach and Landing

The speed definitions for approach and landing conditions are given in Table 3.4. While $V_{MC}$ and $V_{MCG}$ are dedicated to take-off, with a thrust setting at "maximum available take-off power or thrust" – (§25.0149(c)(1)), for landing and approach, similar special definitions are given. The minimum control speed for landing – (§25.0149(f)), $V_{MCL}$, differs from $V_{MC}$ by the engine set to "go-around power or thrust setting on the operating engine(s)" – (§25.0149(f)(6)). Additionally, for aircraft with three or more engines, the minimum control speed for the failure of a second engine, $V_{MCL-2}$, must be established – (§25.0149(g)).

**TABLE 3.4**

**Speed Definitions for Landing according to CS-25/FAR 25 and Corresponding Paragraphs**

| Name | Description | Paragraph |
|---|---|---|
| $V_{MCL}$ | Minimum control speed in landing configuration | §25.0149(f) |
| $V_{MCL-2}$ | Minimum control speed in landing configuration with second engine failure | §25.0149(g) |
| $V_{REF}$ | Reference approach speed | §25.0107(c) |

The reference approach speed, $V_{REF}$, is the minimum airplane steady landing approach speed in the final approach. "A stabilised approach, with a calibrated airspeed of not less than $V_{REF}$, must be maintained down to the 15 m (50 ft) height" – (§25.0125(b)(2)). $V_{REF}$ may not be less than $1.23 \times V_{SR0}$ – (§25.0125(b)(2)(i)(A)), the reference stall speed of the landing configuration, or $V_{MCL}$ – (§25.0125(b)(2)(i)(B)).

### 3.2.3   OPERATING RANGE IN HIGH-LIFT FLIGHT

The definition of minimum and maximum operating speeds prevents the aircraft to be piloted into a non-controllable situation. This, on the counterpart, means that the aerodynamic range of possible flight conditions is not used to the full extent.

Figure 3.3 shows the aerodynamic force polars for a take-off and a landing configuration. In the graphs, the operational limits are defined by the flap extension speed, $V_{FE}$, and the minimum speeds, $V_{2\,min}$ and $V_{REF}$, for take-off and landing, respectively. It is obvious that an aircraft is not operated at the maximum lift coefficient, but of course, the maximum lift coefficient scales the operating range. Additionally, the lift coefficient at zero angle of attack is marked. Due to the lower safety factors for take-off, a typical transport aircraft is operated at positive angles of attack. In landing configuration, the zero angle of attack condition is usually slightly below the center of the operating range.

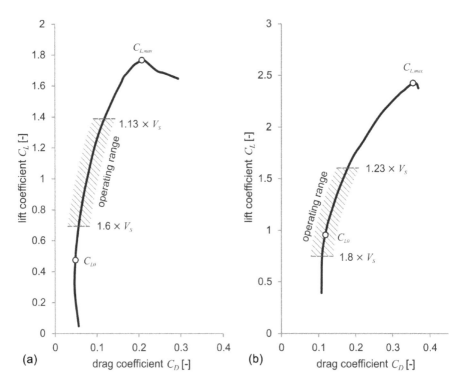

**FIGURE 3.3**   Aircraft force polars for take-off and landing including the indication of the limits of flight operations: (a) take-off; (b) landing.

## 3.3  FLIGHT PHASES

### 3.3.1  TAKE-OFF

The take-off flight path is subdivided into several segments, illustrated in Figure 3.4. "The take-off path extends from a standing start to a point in the take-off at which the aeroplane is 457 m (1500 ft) above the take-off surface, or at which the transition from the take-off to the en-route configuration is completed and $V_{FTO}$ is reached, whichever point is higher" – (§25.0111(a)).

The take-off distance (TOD) from a dry runway is defined as "the horizontal distance along the take-off path from the start of the take-off to the point at which the aeroplane is 11 m (35 ft) above the take-off surface" – (§25.0113(a)). When calculating the TOD, an engine failure of the most critical engine when reaching $V_{EF}$ has to be accounted for – (§25.0111(a)(2)). The TOD is the larger of the demonstrated distance with engine failure and 115% of the one with all engines operating – (§25.0113(a)(2)). Additionally, the TOD has to be established also for wet runway conditions in a similar way – (§25.0113(b)).

At the end of the TOD, the aircraft must have achieved at least the minimum climb speed, $V_{2min}$ – (§25.0111(c)(2)). Up to reaching an altitude of 400 ft above ground, no configuration change is allowed other than retracting the landing gear – (§25.0111(c)(4)). This segment is called the 1st segment climb. The following climb is consecutively called the 2nd segment climb at a speed not less than $V_{2min}$ – (§25.0111(c)(2)) – and reaches up to the end of the take-off path. At this end, the aircraft is accelerated to $V_{FTO}$ and the configuration is changed to the en-route configuration by retracting the high-lift system.

Closely related to the above definitions, the required runway length for take-off has to be established. Besides the determination of the TOD, this requires establishing the accelerate-stop distance (ASD) for aborted take-off. The ASD has to be established according to Figure 3.5 both for all engines operating – (§25.0109(a)(2)) – and for the critical engine failing at $V_{EF}$ – (§25.0109(a)(1)). In the latter case, it is

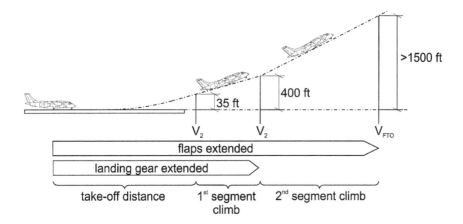

**FIGURE 3.4**  Take-off flight path segments.

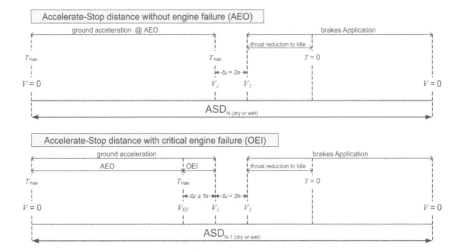

**FIGURE 3.5**  Schematics of the determination of the accelerate-stop-distance according to §25.0109.

assumed that the aircraft accelerates to $V_1$ until the pilot takes first action for aborting the take-off and it is equivalent to at least one second reaction time – (AMC 25.0101(h)(3)). In both cases, an additional 2 seconds time at $V_1$ has to be respected as an additional safety margin.

### 3.3.2  APPROACH AND LANDING

The flight path during approach and landing can be segmented into at least three segments, shown in Figure 3.6. At some height during the flight towards the airport, the aircraft configuration is changed from the en-route configuration to the approach configuration. The final approach consists of "a stabilised approach, with a calibrated airspeed of not less than $V_{REF}$, [...] down to the 15 m (50 ft) height" – (§25.0125(b)(2)). The 50 ft height is the decision height where the pilot has the last decision to proceed

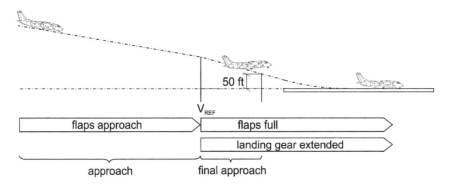

**FIGURE 3.6**  Approach and landing flight path segments.

with the landing or to perform a go-around. In case of a positive decision, the landing distance is "the horizontal distance necessary to land and to come to a complete stop from a point 15 m (50 ft) above the landing surface" – (§25.0125(a)). It is worth to note that the required landing field length of an airport to land the airplane is not defined within Part 25 but within the regulations for air operations, the so-called Part-CAT [6] in Europe and CFR 14 Part 121 [7] in the US for commercial air transportation. Here, it is regulated that the actual landing distance of the aircraft must be less than 60% of the available runway length for turbojet powered airplanes and 70% for turboprop airplanes – (CAT.POL.A.230 in [6], §121.195 in [7]).

## 3.4  FLIGHT PATH SLOPE REQUIREMENTS

For any of the flight phases defined in Section 3.3, the airworthiness requires minimum climb gradients. Throughout the regulations, the climb gradient is noted in percentages

$$\text{climb gradient} = \tan \gamma \cdot 100\%. \tag{3.1}$$

Only for small climb gradients, it is approximately equal to the ratio of rate of climb (vertical speed) and flight speed.

$$\tan \gamma \approx \sin \gamma = \frac{w}{V} \tag{3.2}$$

In general, the required climb gradients are specified depending on the number of engines. The values are given for two, three, and four and more engines. Hereby, the regulations differentiate between normal operations with All Engines Operating (AEO) and cases with One Engine Inoperative (OEI). Additionally, it is differentiated between whether the landing gear is still extended (LGE) or already retracted (LGR).

### 3.4.1  TAKE-OFF CLIMB REQUIREMENTS

Figure 3.7 depicts the required climb gradients for all phases of the take-off path. During take-off with all engines operating, the required climb gradient for the full take-off path (1st and 2nd segment climb) is required to be positive at any point – (§25.0111 (c)(1)). At altitudes above 400 ft (after the landing gear is retracted), a climb gradient of 1.2%/1.5%/1.7% is required depending on the number of engines – (§25.0111 (c)(3)).

The more critical case for climb is engine failure. Therefore, the requirements specify in more detail the required climb gradients for one engine inoperative. Shortly after take-off, when the landing gear is still extended, only a low climb gradient of >0%/0.3%/0.5% is required – (§25.0121 (a)) – that makes it feasible to slightly increase the altitude, while retracting the landing gear and thereby removing a big source of aerodynamic drag. After gear retraction, a significantly higher climb gradient of 2.4%/2.7%/3.0% is required – (§25.0121 (b)). The high climb gradient is

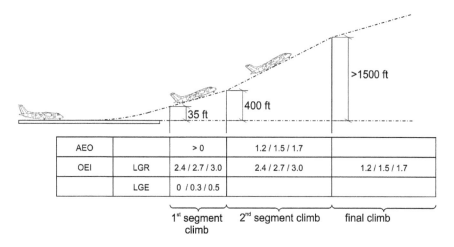

| AEO | | | > 0 | 1.2 / 1.5 / 1.7 | |
|-----|-----|-----|---------------|-----------------|------------------|
| OEI | LGR | | 2.4 / 2.7 / 3.0 | 2.4 / 2.7 / 3.0 | 1.2 / 1.5 / 1.7 |
| | LGE | | 0 / 0.3 / 0.5 | | |

1ˢᵗ segment climb          2ⁿᵈ segment climb          final climb

**FIGURE 3.7**  Climb gradient requirements during take-off.

released for the final climb at the end of the take-off flight to 1.2%/1.5%/1.7%, the same as for all engines operating along the take-off path – (§25.0121 (c)).

At first instance, it seems unnecessary to specify the climb gradient in the 2nd segment climb with all engines operating less than for one engine inoperative. But it has to be remembered that the regulations only require the high-lift configuration not to be changed before reaching an altitude of 400 ft (see 3.3.1). This allows, in case of all engines operating, to select a different setting of the high-lift system, which may result in a lower climb gradient but may be more efficient for aircraft operations. In case of engine failure, such economic reasons are postponed and the safe achievement of a safe flight altitude is pronounced.

### 3.4.2  LANDING CLIMB REQUIREMENTS

Figure 3.8 depicts the required climb gradients in approach or landing configuration. During these flight phases, the requirements are limited to climb operations to perform missed approach or go-around maneuvers. During normal approach and

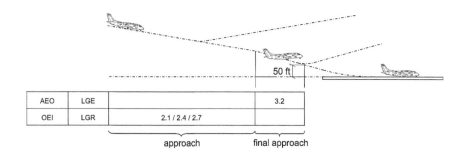

| AEO | LGE | | 3.2 |
|-----|-----|-----------------|-----|
| OEI | LGR | 2.1 / 2.4 / 2.7 | |

approach          final approach

**FIGURE 3.8**  Climb gradient requirements during approach and landing.

landing, no negative climb gradients are specified to achieve the landing approach performance.

When the aircraft approaches, the climb gradient specified for one engine inoperative is 2.1%/2.4%/2.7%, which is slightly less than for take-off – (§25.0121(d)). No further requirement is placed for all engines operating.

For the go-around, that is the landing is aborted during the final approach with flaps in landing position and the landing gear extended, "the steady gradient of climb may not be less than 3·2%, with the engines at the power or thrust that is available 8 seconds after initiation [...]" – (§25.0119). This is probably the highest requirement for climb gradients.

### 3.4.3 Steep Approach

With amendment 13 to the CS-25 [5], special airworthiness regulations were formulated for steep approach capabilities of an aircraft. It should be highlighted that the FAR 25 is not extended by such regulations[1]. These regulations have been specially integrated to certify the operability of aircraft at the airport in London City [9]. At this airport, the approach glide path is increased from the usual 3° ILS glide path to a 5.5° glide path. And, it is the only airport requiring the steep approach capabilities to be "contained in its Flight Manual data and procedures for approach path angles of 5.5° or steeper" – (AIP EGLC AD 2.20, 1(a)). As the Aircraft Flight Manual is a document approved by the certification, respective rules needed to be elaborated.

The certification requirements for Steep Approach Landing (SAL) capabilities are compiled in the CS 25 Appendix Q. The climb gradient required for the specific steep approach configuration is equivalent to the one of normal approach with 2.1%/2.4%/2.7% – (§CS-Q(SAL) 25.4). Additionally, "it must be possible to achieve an approach path angle 2° steeper than the selected approach path angle in all configurations" – (§CS-Q(SAL) 25.5(d)) – used during approach and landing.

Turboprop aircraft are most capable to achieve the steep approach requirements. Only few turbojet aircraft are certified for the steep approach capabilities, mainly, regional jets as the British Aerospace BAe 146, Embraer ERJ135, Fokker 70, and Dornier 328-300 with capacities between 35 and 100 passengers. Larger and newer aircraft certified for the steep approach are Airbus A318, Bombardier C-Series (now Airbus A220), and Embraer E-Jet series.

## 3.5  LOAD CASES FOR STRESS ASSESSMENT

A basic intention of the certification specifications is to give agreed design rules for the structural dimensioning of the overall aircraft and all its components. For this, the regulations prescribe the envelope of structural loads that must be substantiated during aircraft design. Within this section, only the essential requirements affecting the high-lift system and originating from aerodynamic forces are discussed.

Aerodynamic loads have to be established for the full range of aircraft operations. This does not only mean that the loads must cover the speed range but also the mass range of the aircraft as the aerodynamic forces, especially in lift direction, scale with it. Therefore, loads assumptions for high-lift devices have to cover the full mass range from Operating Empty Weight (OEW) to Maximum Take-Off Weight (MTOW), or Maximum Landing Weight (MLW), respectively.

The envelope of flight with high-lift systems deflected has already been discussed in Chapter 1 to motivate the need of high-lift systems. At that stage, the speed range of the envelope has already been sketched in Figure 1.6. From the regulations regarding the flight speeds in Section 3.2, the speed range for the structural design has been defined more precisely between the 1-g stall speed, $V_{S1g}$, and the flap design speed, $V_F$, for every configuration. Since the actual force is always related to speed and the aerodynamic coefficients, in principle, subdivided into the aircraft components, the regulations specify the load's conditions over the full range of the flight's envelope.

The loads to be respected are given in terms of load factors. "Flight load factors represent the ratio of the aerodynamic force component (acting normal to the assumed longitudinal axis of the aeroplane) to the weight of the aeroplane. A positive load factor is one in which the aerodynamic force acts upward with respect to the aeroplane" – (§25.0321(a)).

### 3.5.1 MANEUVER LOADS

A maneuver load has to be considered as a quasi-static procedure where the acting forces result in the change of the aircraft's motion. Maneuvers in this sense include, e.g., pitch-up and pitch-down maneuvers to change the attitude of the aircraft as well as turning flight. The major characteristic to be respected is the additional load factor $n_z$ that multiplies the aircraft 1-g weight to compensate centrifugal forces during the curved flight motion.

The load factor directly relates to the flight speed and the radius of the curved flight path. Figure 3.9 shows the conditions for a pitch-up maneuver. The load factor relates to the radius of the flight path by at maximum

$$n_{z,\max} = 1 + \left(V^2 \middle/ Rg\right). \tag{3.3}$$

Note that the maximum value is at the lowest point of the flight path. For a pitch-down, eq. (3.3) is applied with a negative radius.

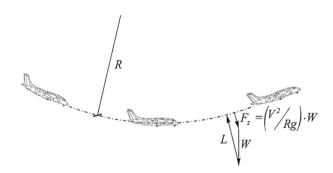

**FIGURE 3.9**   Additional loads during pitch-up maneuver.

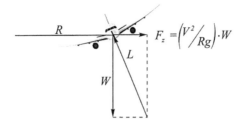

**FIGURE 3.10** Additional loads during turn maneuver.

For turning flight, the radius of the flight path corresponds to the velocity and a corresponding bank angle of the aircraft. As Figure 3.10 shows, the aircraft has to produce an additional lift and has to be rolled so that the additional portion of the lift has the necessary component towards the center of the curved flight path to compensate the centrifugal force.

$$n_z = \sqrt{1+\left(V^2/Rg\right)^2}. \tag{3.4}$$

The corresponding bank angle is given by

$$\tan\phi = V^2/Rg \tag{3.5}$$

which makes it also possible to directly relate the load factor to the bank angle during turning flight

$$n_z = \sqrt{1+\left(\tan\phi\right)^2}. \tag{3.6}$$

The certification specifications required for an aircraft with high-lift devices deployed sustaining "manoeuvring to a positive limit load factor of 2.0" – (§25.0345(a)(1)). This formulation implies that no negative load factors have to be accounted for the clean configuration and the lowest load factor to be respected is $n_{z,min} = 0$. For the landing configuration, in addition to the load envelope up to $n_{z,max} = 2$ at MLW, it is required to respect for load factors up to $n_z = 1.5$ for weights up to MTOW – (§25.0345(d)). This is intended to allow for overweight landings and to remove the necessity to drop fuel in case of immediate returns to the airport.

### 3.5.2 GUST LOADS

While for the cruise flight, a large set of characteristic gusts have to be assessed – (§25.0335), the required analysis for high-lift systems is limited "to positive and negative gusts of 7.62 m/sec (25 ft/sec) EAS acting normal to the flight path in level flight" – (§25.0345(a)(2)) – and "a head-on gust of 7.62 m/sec (25 fps) velocity

(EAS)" – (§25.0345(b)(2)). The shape of the gust is uniquely specified to follow the gust profile

$$U_{gust}(s) = \frac{U_{ds}}{2}\left(1 - \cos\frac{\pi s}{12.5c}\right); 0 < s < 25c \tag{3.7}$$

where $s$ is the distance penetrated into the gust – (§25.0335(a)(2)). "The analysis must take into account the unsteady aerodynamic characteristics and rigid body motions of the aircraft" – (§25.0345(a)(2)). The latter is essential, as the evasive motion of the aircraft, on the one hand, reduces the maximum load factor during the gust by up to 60% but introduces a second load factor with opposite orientation on the phase when the gust has vanished and the aircraft returns to steady flight.

The vertical acceleration of the aircraft is depending on the actual speed, horizontal and vertical speed depending on the orientation of the gust.

$$\begin{pmatrix} V_{gust} \\ w_{gust} \end{pmatrix}(t) = \frac{25c}{V_\infty} \cdot \alpha_{gust} \cdot U_{gust}(s)$$
$$V_{eff}(t) = V_\infty + V_{gust}(t) \tag{3.8}$$
$$w_{eff}(t) = w_{gust}(t) - w(t).$$

Additionally, the lift is depending on the actual angle of attack.

$$\alpha_{eff}(t) = \alpha + \tan^{-1}\left(\frac{w_{eff}(t)}{V_{eff}(t)}\right). \tag{3.9}$$

The vertical acceleration is then calculated by

$$a(t) = \frac{C_L|_{\alpha_{eff}(t)}\frac{\rho}{2}\left(V_{eff}^2(t) + w_{eff}^2(t)\right)A_{ref}}{m} - g. \tag{3.10}$$

The rigid body motion is finally given by the differential equation

$$\frac{dw(t)}{dt} = \frac{\rho}{2}\frac{A_{ref}}{m}C_L|_{\alpha_{eff}(t)}\left(\left(V_\infty + V_{gust}(t)\right)^2 + \left(w_{gust}(t) - w(t)\right)^2\right) - g, \tag{3.11}$$

which can be solved by time integration.

### 3.5.3 OPERATING LOADS

Other than for the cruise condition, the high-lift system faces additional load cases during operation. Operation in this sense means the change of the configuration from one setting to another. It is required that "the lift device control must be designed to retract the surfaces from the fully extended position, during steady flight at maximum continuous engine power at any speed below $V_F$ + 17 km/hr (9.0 knots)" – (§25.0697(d)). The speed is defined higher so that the retraction of the high-lift devices is even possible at speeds exceeding the flap design speed $V_F$ which would impose a structural overload. This load condition, therefore, does not impact the structural design of the high-lift device itself but for the system used for its deflection.

## 3.6   NOISE CLASSIFICATION

The requirements for noise of an aircraft are included in CS-36 [10]. The (full) regulations of CS-36 read: "The aircraft must be designed to comply with the applicable noise requirements defined under 21.B.85(a)" – (CS-36, §36.1).

§21.B.85(a) refers to the EASA Part 21 [11]:

> The Agency shall designate and notify to the applicant for a type-certificate or restricted type-certificate for an aircraft, for a supplemental type-certificate or for a major change to a type-certificate or to a supplemental type-certificate, the applicable noise requirements established in Annex 16 to the Chicago Convention, Volume I, Part II, Chapter 1 and:
>
>    1. for subsonic jet aeroplanes, in Chapters 2, 3, 4 and 14;
>    2. for propeller-driven aeroplanes in Chapters 3, 4, 5, 6, 10, and 14;
>    3. for helicopters, in Chapters 8 and 11;
>    4. for supersonic aeroplanes, in Chapter 12; and
>    5. for tilt rotors, in Chapter 13.

**(EASA Part 21, 21.B.85 (a))**

In the United States, a similar regulation is CFR 14 Part 36 (FAR 36) [12]. In contrast to the CS-36, the FAR 36 differentiates between different aircraft "stages." While the different stages introduce slightly different noise levels for other – mainly older – types of aircraft, starting from "Stage 4," the FAR 36 directly references to the so-called ICAO Annex 16 [13] – (FAR 36, §B36.5(d), and §B36.5(e)). It specifies target noise levels as well as the measurement procedure. In the following, we concentrate on subsonic jet airplanes, which are specified in ICAO Annex 16, Chapter 3. Noise level limits are specified for three different locations: fly-over, lateral, and approach.

The fly-over noise measurement is related to take-off. Figure 3.11 illustrates the corresponding measurement point that is located 6500 m past the start-of-roll in the extension of the runway centerline – (ICAO Annex 16, Ch. 3, §3.3.1(b)). The dominant noise source during take-off is for sure the engine itself. Nevertheless, aerodynamics impacts the fly-over noise directly, as the perceived noise on the ground is a function of the distance of the noise source to the observer by the square of the sound pressure level scales with the inverse of the distance.

$$p^2 \sim \frac{1}{h} \qquad\qquad (3.12)$$

The altitude at the fly-over point depends on the length of the ground-roll, where the maximum lift coefficient is an important factor as it scales the lift-off speed, $V_{LOF}$,

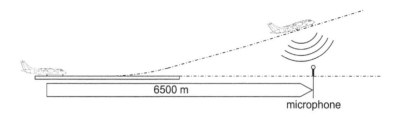

**FIGURE 3.11**  Fly-over noise measurement.

and the climb performance in terms of the climb gradient, which is a function of the lift to drag ratio.

The lateral noise is measured for jet-driven aircraft "at a point on a line parallel to and 450 m from the runway centre line, where the noise level is a maximum during take-off" – (ICAO Annex 16, Ch. 3, §3.3.1(a)(1)). For propeller-driven aircraft, the lateral noise can alternatively also be measured along the centerline at a point where the aircraft achieves an altitude above ground of 650 m – (ICAO Annex 16, Ch. 3, §3.3.1(a)(2)). The lateral noise is a pure function of the engine noise as the distance to the microphone is kept constant. Therefore, aerodynamics is not influencing this.

The approach noise is measured in the last part of the approach with engines on idle. As illustrated in Figure 3.12, the measurement point is placed "on the extended centre line of the runway 2 000 m from the threshold. On ground level this corresponds to a position 120 m (394 ft) vertically below the 3° descent path originating from a point 300 m beyond the threshold" – (ICAO Annex 16, Ch. 3, §3.3.1(c)). Since the approach path is therefore given, the distance to the microphone is not affected by aerodynamics. But now, the engines are no longer the dominant noise source but the airframe. As the airframe noise scales with the flight speed, the perceived noise can be reduced by a reduced approach flight speed that is again directly related to the maximum lift coefficient.

The maximum noise levels specified in ICAO Annex 16 are defined as weight dependent. Above a threshold weight value, the maximum noise level is constant. Down from this aircraft mass, the maximum noise level reduces with the logarithm of the weight until reaching a minimum noise limit or a lower weight threshold. The noise scale used is the so-called effective perceived noise level, EPNL – (ICAO Annex 16, Appendix 2, Section 4), that combines a weighting of the maximum noise level during the fly-over and the duration of noise perception and is measured in effective perceived noise decibel, EPNdB.

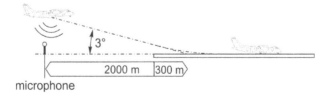

**FIGURE 3.12**  Approach noise measurement and influence factors.

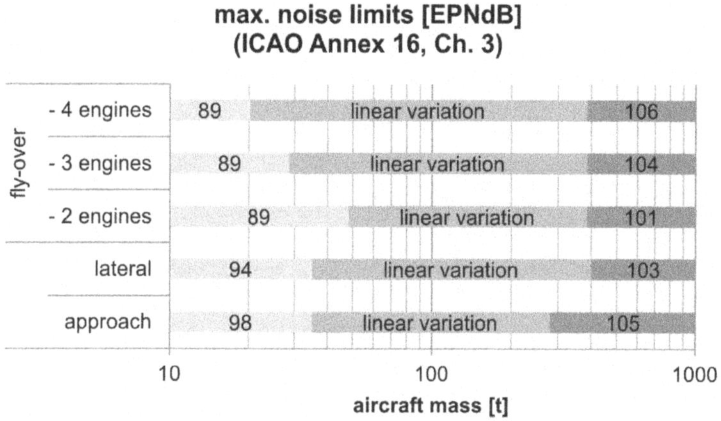

## max. noise limits [EPNdB]
## (ICAO Annex 16, Ch. 3)

FIGURE 3.13   ICAO Annex 16, Chapter 3 noise limits depending on aircraft mass.

Figure 3.13 summarizes the maximum noise levels at the threshold weights for all three measurement directions as defined in Chapter 3 for all subsonic jet aircraft certified after 6 October 1977 and before 1 January 2006. For the fly-over noise during take-off, the upper threshold is specified at an aircraft mass of 385 t and with a limit value depending on the number of engines. The noise level reduces linearly by a rate of -4 EPNdB for every halving of the aircraft mass until reaching a noise level of 89 EPNdB. Aircraft are not required to emit less noise than this independent of weight. The lateral noise upper threshold is at 105 EPNdB for aircraft masses exceeding 400 t independent of the number of engines and a lower threshold of 94 EPNdB for aircraft masses at or below 35 t with a linear variation with the logarithm of mass in between. The approach noise is limited by 105 EPNdB for aircraft masses at or above 280 t and a lower limit of 98 EPNdB for aircraft masses at or below 35 t with the same linear variation with the logarithm of the mass in between.

For aircrafts certified past 1 January 2006 but before 31 December 2017, more restrictive noise limits were established within Chapter 4. Although the noise limit levels were unchanged, an additional requirement was formulated that the cumulative noise level, which is the sum of the lateral, fly-over, and approach noise levels, has to be 10 EPNdB less than the cumulative sum of the noise level limits according to Chapter 3. Additionally, each sum of two noise measurement points must be at least 2 EPNdB less than the sum of the corresponding noise level limits.

For aircrafts certified past 31 December 2017, the newly established Chapter 14 sharpens the requirement on the cumulative noise level to 17 EPNdB less than the cumulative sum of the noise level limits according to Chapter 3. Now, at each noise measurement point, the noise level must be at least 1 EPNdB less than the corresponding noise level limits, by this reducing the effective noise level limits by 1 EPNdB with respect to Chapter 3. In addition, Chapter 14 introduces additional lower noise limits for aircraft with less than 8,618 kg of maximum take-off weight.

Figure 3.14 summarizes the development of the cumulative noise level limit depending on the date of certification of the aircraft for turbojet aircraft.

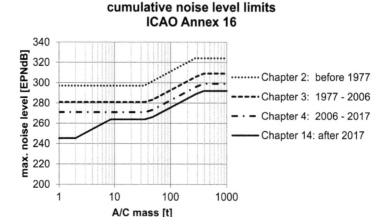

**FIGURE 3.14**  Cumulative noise level limits for aircraft of different ages (certification date) according to ICAO Annex 16.

## NOTE

1. There is no certification standard for steep approach landing in the United States. Only the Advisory Cicrcular AC 120-29A [8] gives some hints how to proceed.

## REFERENCES

[1] Federal Aviation Administration (2018) Code of Federal Regulations Title 14 Part 25 (Annual Edition), Amendment 146,25 – US Government Publishing Office.
[2] European Aviation Safety Agency (2020) Certification Specifications and Acceptable Means of Compliance for Large Aeroplanes, CS-25, Amendment 26.
[3] Federal Aviation Administration (2002) Code of Federal Regulations Title 14 Parts 1, 25, 36, and 97, Doc. No. 28404, 67 FR 70825, Nov. 26, 2002, as amended by Amdt. 25–108.
[4] Federal Aviation Administration (1964) Code of Federal Regulations Title 14 Parts 1, 4a, 4b, 25, and 91, Doc. No. 5066; Amendment Nos. 1–6, 4a–0, 4b–0, 25–0, 91–10.
[5] European Aviation Safety Agency (2013), Certification Specifications and Acceptable Means of Compliance for Large Aeroplanes, CS-25, Amendment 13.
[6] European Aviation Safety Agency (2020) EU regulation No 965/2012, Annex IV (Part-CAT).
[7] Federal Aviation Administration (2021) Code of Federal Regulations Title 14 Part 121, US Government Publishing Office.
[8] Federal Aviation Administration (2002) Criteria for Approval of Category I and Category II Weather Minima for Approach, Advisory Circular 120–29A.
[9] Civil Aviation Authority (2016) United Kingdom AIP, ELGC – London City, AD 2. EGLC-8.
[10] European Aviation Safety Agency (2019), Certification Specifications and Acceptable Means of Compliance and Guidance Material for Aircraft Noise, CS-36, Amendment 5.

[11] European Commission (2016) Part 21 – Certification of Aircraft and Related Products, Parts and Appliances, and of Design and Production Organisations, Annex to Commission Regulation (EC) No 1702/2003, Issue 2, Amendment 6.

[12] Federal Aviation Administration (2005) Code of Federal Regulations Title 14 Part 36 (Annual Edition), US Government Publishing Office.

[13] International Civil Aviation Organization (2017) Environmental Protection, Annex 16 to the Convention on International Civil Aviation, Volume I, Aircraft Noise, Eight Edition, July 2017, Amendment 13, effective January 1, 2018.

# 4 Aircraft Performance at Take-Off and Landing

## NOMENCLATURE

| | | | | | | |
|---|---|---|---|---|---|---|
| $a$ | $m/s^2$ | Acceleration | $s_A$ | $m$ | Approach distance |
| $a_b$ | $m/s^2$ | Braking deceleration | $s_{FL}$ | $m$ | Flare distance |
| $a_{FL}$ | $m/s^2$ | Flare deceleration | $s_{FR}$ | $m$ | Free roll distance |
| $a_{GR}$ | $m/s^2$ | Ground roll acceleration | $s_{GR}$ | $m$ | Ground roll distance |
| $a_{LOF}$ | $m/s^2$ | Lift-off acceleration | $s_{LOF}$ | $m$ | Lift-off distance |
| $A_{ref}$ | $m^2$ | Reference wing area | $s_{IC}$ | $m$ | Initial climb distance |
| $b$ | $m$ | Wing span | $s_R$ | $m$ | Level out distance |
| $c$ | $m$ | Chord | $T$ | $N$ | Thrust |
| $c_L$ | $-$ | Sectional lift coefficient | $T_S$ | $N$ | Static thrust |
| $c_L'$ | $-$ | Sectional lift coefficient in ground effect | $U$ | $m/s$ | Velocity |
| $C_D$ | $-$ | Drag coefficient | $V$ | $m/s$ | Flight speed |
| $C_D'$ | $-$ | Drag coefficient in ground effect | $V_{LOF}$ | $m/s$ | Lift-off speed |
| $C_{D,i}$ | $-$ | Induced drag coefficient | $V_R$ | $m/s$ | Rotation speed |
| $C_{D0}$ | $-$ | Zero drag coefficient | $V_{REF}$ | $m/s$ | Reference approach speed |
| $C_{D,SP}'$ | $-$ | Drag coefficient in ground effect with speed brakes deflected | $V_{SR}$ | $m/s$ | Reference stall speed |
| $C_L$ | $-$ | Lift coefficient | $V_{TD}$ | $m/s$ | Touch-down speed |
| $C_L'$ | $-$ | Lift coefficient in ground effect | $V_2$ | $m/s$ | Climb speed |
| $C_{L0}$ | $-$ | Zero lift coefficient | $w$ | $m/s$ | Climb/sink rate |
| $C_{L0}'$ | $-$ | Zero lift coefficient in ground effect | $W$ | $kg$ | Weight |
| $C_{L0,SP}'$ | $-$ | Zero lift coefficient in ground effect with speed brakes deflected | $\alpha$ | $°$ | Angle of attack |
| $C_{L,max}$ | $-$ | Maximum lift coefficient | $\alpha_i$ | $°$ | Induced angle of attack |
| $C_{L,max}'$ | $-$ | Maximum lift coefficient in ground effect | $\alpha_{LOF}$ | $°$ | Angle of attack at lift-off speed |
| $D$ | $N$ | Drag force | $\alpha_{max}$ | $°$ | Maximum rotation angle on ground |
| $\varepsilon_E$ | $°$ | Engine axis incidence andle | $\beta$ | $-$ | Torenbeek lift impact coefficient |
| $e$ | $-$ | Oswald's efficiency factor | $\beta_D$ | $-$ | Phillips & Hunsaker lift impact coefficient on drag |
| $F_{eff}$ | $N$ | Effective acceleration force | $\beta_L$ | $-$ | Phillips & Hunsaker lift impact coefficient on lift |
| $g$ | $m/s^2$ | Gravity acceleration | $\gamma$ | $°$ | Flight path angle |
| $h$ | $m$ | Height above ground | $\Gamma'$ | $m^2/s$ | Circulation in ground effect |

DOI: 10.1201/9781003220459-4

| | | | | | |
|---|---|---|---|---|---|
| $h_f$ | $m$ | Flare initiation height | $\delta_D$ | – | Phillips & Hunsaker coefficient on drag |
| $k$ | – | Pistolesi circulation influence factor | $\delta_L$ | – | Phillips & Hunsaker coefficient on lift |
| $L$ | $N$ | Lift force | $\Lambda$ | – | Aspect ratio |
| $m$ | $kg$ | Mass | $\Lambda'$ | – | Effective aspect ratio in ground effect |
| $n_z$ | – | Load factor | $\mu_R$ | – | Roll friction coefficient |
| $P$ | $W$ | Power | $\mu_{R,b}$ | – | Roll friction coefficient with brakes |
| $r$ | – | Pistolesi velocity influence factor | $\rho$ | $kg/m^3$ | Density |
| $R$ | $m$ | Flight path radius | $\sigma$ | – | Induction coefficient |
| $s$ | $m$ | Distance | | | |

After having addressed the potential of lift generation in viscous compressible flow and flight safety requirements for aircraft, the next question is on benefits in terms of aircraft performance. In general, high-lift systems are in use only during a short period of an overall flight mission. Nevertheless, a high-lift system has to be designed to make best use of it in all low-speed flight phases, aside from reducing the flight speed.

Aircraft manufacturers are not very open to communicating the benefits of high-lift systems accompanied by numbers. One of the few examples given in literature is provided by Meredith [1]. He outlines some impressive relations how an improved high-lift system can improve the whole aircraft, as there are:

1. A 0.10 increase in lift coefficient at constant angle of attack is equivalent to reducing the approach attitude by about one degree. For a given aft body-to-ground clearance angle, the landing gear may be shortened resulting in a weight savings of 1400 Ib [635 kg].
2. A 1.5% increase in maximum lift coefficient is equivalent to a 6600 lb [2994 kg] increase in payload at a fixed approach speed.
3. A 1% increase in take-off L/D is equivalent to a 2800 Ib [1270 kg] increase in payload or a 150 nm [278 km] increase in range.

**(Meredith [1] p. 19-1)**

Other research has shown that a flight path altitude increased by 100 m due to improved climb or steep descent reduces the perceived noise pressure level on ground by 1 dB. It is the intention of this chapter to outline which aerodynamic parameters determine the performance of the high-lift system in the overall aircraft context.

## 4.1   GENERAL RELATIONS

The performance of an aircraft in steady state flight is related to the balance of lift and weight, and of drag and thrust. Figure 4.1 shows the forces acting on the aircraft. For a non-accelerated flight, the balance of forces in the aerodynamic system is given by

$$W \cos \gamma = L + T \sin(\alpha + \varepsilon_E)$$
$$T \cos(\alpha + \varepsilon_E) = D + W \sin \gamma, \tag{4.1}$$

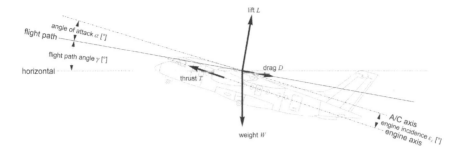

**FIGURE 4.1**  Balance of forces of an aircraft in steady state flight.

where the thrust axis is inclined to the aircraft axis by $\varepsilon_E$. Dividing the two parts of eq. (4.1) leads to the determination of the flight path angle, or climb gradient

$$\tan\gamma = \frac{T\cos(\alpha+\varepsilon_E)-D}{T\sin(\alpha+\varepsilon_E)+L} \tag{4.2}$$

and the lift needed to carry the aircraft weight is derived to

$$L = W\cdot\cos\gamma - T\cdot\sin(\alpha+\varepsilon_E). \tag{4.3}$$

As lift and drag are themselves functions of the angle of attack, these relations can only be solved iteratively.

In a first approximation, the installation angle of the engine with regard to the aircraft axis can be neglected. In a second common approximation for low flight path angles and low angles of attack, the aircraft weight approximately equals the lift force

$$W \approx L. \tag{4.4}$$

The relation eq. 4.2 can then be rearranged to give the **flight path angle**

$$\sin\gamma \approx \frac{T}{W}\cos\alpha - \frac{C_D}{C_L}. \tag{4.5}$$

In this relation, the aircraft configuration part in terms of the thrust-to-weight ratio is separated from the aerodynamic part in terms of the inverse of the lift-to-drag ratio. Nevertheless, as the lift together with the thrust is needed to compensate the aircraft weight, the corresponding lift-to-drag ratio has to comply with the corresponding flight speed and angle of attack.

By the velocity triangle, the **climb or sink rate** relates to the flight path angle and the flow velocity by

$$\sin\gamma = \frac{V}{w} \Rightarrow w = V\sin\gamma. \tag{4.6}$$

With the use of eqs. (1.2) and (4.5), it can be expressed by

$$w \approx \frac{T \cdot V}{W} \cos\alpha - \sqrt{\frac{2}{\rho_\infty} \frac{W}{A_{ref}} \frac{C_D}{C_L^{3/2}}} \tag{4.7}$$

or

$$w \approx \sqrt{\frac{2}{\rho_\infty} \frac{W}{A_{ref}}} \left( \frac{T}{W} \frac{1}{C_L^{1/2}} \cos\alpha - \frac{C_D}{C_L^{3/2}} \right). \tag{4.8}$$

Therefore, the sink rate not only depends on the thrust-to-weight ratio but also on the wing loading. The aerodynamic part is now described by the climb index $C_L^{3/2}/C_D$. The type of propulsion system determines the adequate equation. For propeller-driven aircraft, usually the propulsion power

$$P = T \cdot V \tag{4.9}$$

is assumed to be constant with respect to the flight speed. This recommends the use of eq. (4.7). For jet-powered aircraft, the thrust is assumed to be constant with respect to the flight speed, leading to eq. (4.8).

It has to be highlighted here, that, especially for take-off, the simplification of low flight path angles and low angles of attack are not truly valid. Especially during take-off, the high climb gradient results in a major contribution of the thrust to compensate for the aircraft weight. Therefore, the lift coefficient in this flight phase is much less and the aircraft probably operates at a different lift-to-drag ratio. For an exact determination of the take-off climb path, the iterative solution of eq. (4.2) and (4.3) is necessary.

The propulsion thrust or power for an aircraft is almost constant, although depending on environmental conditions such as temperature and pressure (altitude). In contrast, weight is not. It depends on the actual payload, fuel, and other operational aspects. For aircraft design, several specific weights are defined to specify the weight condition for several design aspects. The different weight definitions are listed in Table 4.1. For the design of high-lift devices of an aircraft, the most important are the upper limits of the associated weights. Derived from eqs. (1.2) and (4.4) at a maximum weight, either the target of the necessary lift coefficient to fly at a certain speed or the speed associated with a certain lift coefficient is maximum. Therefore, during the aircraft design process, two specific weight conditions are defined, which are:

1. The maximum take-off weight (MTOW) is the highest allowable weight for the aircraft to take-off.
2. The maximum landing weight (MLW) is the highest allowable weight for the aircraft to land.

In an early stage of aircraft development, the aircraft-related values might not be frozen. The situation is even worse if developing high-lift airfoils even independent of a distinct aircraft. In consequence, developing a high-lift airfoil for good climb

## TABLE 4.1
## Weight Condition Definitions of a Transport Aircraft

| Abbrev. | Weight Condition | Definition |
|---------|------------------|------------|
| MEW | Manufacturer's empty weight | Structural weight of the aircraft including all integral components (oil for hydraulics, brakes) |
| OEW | Operational empty weight | + Equipment, crew, lubricants (water, engine oil) |
| DOW | Dry operating weight | + Mission equipment (e.g., catering, newspapers) |
| ZFW | Zero fuel weight | + Payload |
| LW | Landing weight | + Reserve fuel |
| TOW | Take-off weight | + Mission fuel |

performance is, at first glance, not possible without knowing the aircraft details. On the second view, most transport aircraft in service look similar from a global perspective, being a tube, a wing, some engines, and an empennage. It is therefore worth to analyze typical values of the thrust-to-weight ratio or wing loading for a transport aircraft. In [2], data of weight and propulsion is given for most aircraft in service end of 20th century. Figure 4.2 shows a summary of the data in terms of maximum take-off weight and static thrust. It can be seen that the correlation of both values is nearly linear, independent of the size of the aircraft. A regression analysis (dashed line in Figure 4.2) delivers

$$\frac{T_S}{MTOW \cdot g} = 0.271 \pm 0.010 \tag{4.10}$$

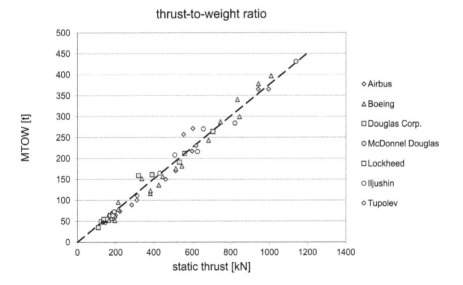

**FIGURE 4.2** Maximum take-off weight versus static thrust for civil transport aircrafts in operation end of 20th century.

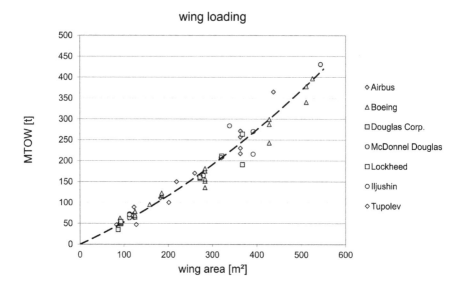

**FIGURE 4.3**   Maximum take-off weight versus wing area for civil transport aircraft in operation end of 20th century.

as an approximation for the thrust-to-weight ratio regarding the maximum take-off weight. Not shown, similar analysis for the maximum landing weight and the operational empty weight result in the regression approximations

$$\frac{T_S}{MLW \cdot g} = 0.368 \pm 0.015$$

$$\frac{T_S}{OEW \cdot g} = 0.566 \pm 0.037.$$

(4.11)

Figure 4.3 shows a similar correlation of the maximum take-off weight with respect to wing area. Here, the correlation is not linear and the spreading is larger. The regression for the MTOW and the other relevant weights is calculated to

$$\frac{MTOW}{A_{ref}} = 480.28 + 0.514 \cdot A_{ref}$$

$$\frac{MLW}{A_{ref}} = 453.23 + 0.131 \cdot A_{ref}$$

$$\frac{OEW}{A_{ref}} = 263.66 + 0.163 \cdot A_{ref}.$$

(4.12)

### 4.1.1   PHASES OF HIGH-LIFT FLIGHT

Before going into the specifics of high-lift system aerodynamic performance, it is worth to understand in which phases of flight they are used. Figure 4.4 shows the different flight phases with deflected high-lift systems.

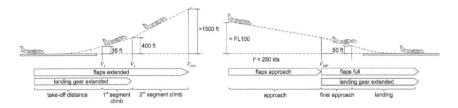

**FIGURE 4.4** Phases of flight with high-lift systems in use.

Beginning at the start of the journey, the lift-off is the point when the aircraft has been accelerated, left ground, and achieved a safety altitude of 35 ft. From this point, the 1st segment climb runs until an altitude of 400 ft where the gear is retracted. The most important climb segment is the following 2nd segment climb that continues until the aircraft is accelerated again and the high-lift devices are retracted. Regulations require that this segment does not end below 1500 ft above ground. But it strongly depends on the aerodynamic efficiency of the high-lift system if it is worth to keep them deflected up to higher altitudes.

Depending on the flight procedure and aircraft characteristics, the high-lift system is deflected not before passing the flight level FL100 (pressure altitude 10,000 ft above QNH 1013.25 hPa) and mainly to achieve the maximum flight speed of 250 kts mentioned in Section 1.2. The approach phase takes until the aircraft extends the landing gear and the high-lift system to the landing setting and achieves a stabilized final approach glide path given by the air traffic control, which is most commonly a glide path angle of $\gamma = 3°$. In the landing phase, the aircraft is kept in this configuration up to an altitude of 50 ft above ground. This is the decision height where the pilot at latest takes the choice to land and touch-down or to perform a go-around.

### 4.1.2 GROUND EFFECT

Important effects to respect for analyzing aircraft performance in the proximity of ground are the changes of lift and drag coefficients. The estimation of the so-called ground effect is possible by potential theory. By mirroring the flow field at the ground surface, a symmetry condition is obtained that prescribes the main feature of the ground: The flow is not going through it.

Figure 4.5 illustrates that the mirrored circulation induces a flow velocity in upward direction that has to be compensated by an increased circulation of the lifting body itself. Basic analysis using lifting line theory by Wieselsberger [3] indicated that the mirrored flow field acts like an effective change in wing aspect ratio leading to a negative induced incidence

$$\Delta\alpha_i = -\sigma \frac{C_L}{\pi\Lambda} \tag{4.13}$$

and an according reduction of the induced drag by

$$\Delta C_{D,i} = -\sigma \frac{C_L^2}{\pi\Lambda}. \tag{4.14}$$

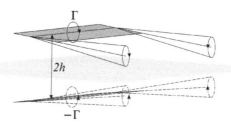

**FIGURE 4.5**   Estimation of ground effect by potential theory by mirroring of the flow field.

The induction coefficient σ for a wing with elliptical circulation distribution was obtained by Prandtl [4] from biplane analysis

$$\sigma = \frac{1-0.66\dfrac{2h}{b}}{1.05+3.7\dfrac{2h}{b}}; \; \tfrac{1}{15} < \frac{2h}{b} < \tfrac{1}{2} \tag{4.15}$$

**where $h$ is the distance of the circulation center to the ground**[1]. In the context of drag, the dominant characteristic value is therefore the ratio of altitude to wing span. The previous formulas assume an elliptic lift distribution. Since the ground effect acts on the lift induced drag coefficient only, the total drag coefficient in ground proximity is given by

$$C_D' = C_{D0} + \frac{C_L^2}{e\pi\Lambda'}. \tag{4.16}$$

The influence is also described as a change of the effective aspect ratio, which can be applied in cases where the lift distribution is not elliptical, too. The according ratio of the effective aspect ratio is derived from the Prandtl factor eq. (4.15) by

$$\frac{\Lambda}{\Lambda'} = 1-\sigma. \tag{4.17}$$

Another expression for the effective aspect ratio in ground effect is given by McCormick [5][2]

$$\frac{\Lambda}{\Lambda'} = \frac{16\left(\dfrac{h}{b}\right)^2}{1+16\left(\dfrac{h}{b}\right)^2} \tag{4.18}$$

based on analysis of a simple mirrored vortex system. Suh & Ostowari [6] derive for the same single vortex

$$\frac{\Lambda}{\Lambda'} = 1 - \frac{2e}{\pi^2}\ln\left(1+\left(\pi\frac{b}{4h}\right)^2\right). \tag{4.19}$$

Torenbeek [7] gives a fit to lifting-line. He is the first to realize that the ground effect on drag is not only related to the height above ground but also to aspect ratio and lift coefficient[3].

$$\frac{\Lambda}{\Lambda'} = \frac{1 - e^{\left[-2.48\left(\frac{2h}{b}\right)^{0.768}\right]}}{1 - \frac{\beta C_L}{4\pi\Lambda\left(\frac{h}{c}\right)}}$$

$$\beta = \sqrt{1 + \left(\frac{2h}{b}\right)^2} - \left(\frac{2h}{b}\right).$$

(4.20)

Phillips & Hunsaker [8] refine the Torenbeek equation to fit to lifting-line solutions. Like Torenbeek, they acknowledge the influence of the wing aspect ratio and lift coefficient, but recognize also the influence of wing taper ratio, so they formulate the approximation of the ground effect to

$$\frac{\Lambda}{\Lambda'} = \left(1 - \delta_D e^{\left[-4.74\left(\frac{h}{b}\right)^{0.814}\right]} - \left(\frac{h}{b}\right)^2 e^{\left[-3.88\left(\frac{h}{b}\right)^{0.758}\right]}\right) \Big/ \beta_D$$

$$\delta_D = 1 - 0.157\left(\lambda^{0.775} - 0.373\right)\left(\Lambda^{0.417} - 1.27\right)$$

$$\beta_D = 1 - \frac{0.0361 C_L^{1.21}}{\Lambda^{1.19}\left(\frac{h}{b}\right)^{1.51}}.$$

(4.21)

Hoerner & Borst [9] relate the effective aspect ratio to the change of induced angle of attack

$$\frac{\Lambda}{\Lambda'} = \frac{\left(d\alpha_i/dC_L\right)'}{d\alpha_i/dC_L}.$$

(4.22)

They give a formula for the change of the induced angle of attack gradient[4]

$$\frac{\left(d\alpha_i/dC_L\right)'}{d\alpha_i/dC_L} = \frac{33\left(\frac{h}{b}\right)^{3/2}}{33\left(\frac{h}{b}\right)^{3/2} + 1}.$$

(4.23)

Figure 4.6 compares the different approaches to calculate the ground effect on the drag coefficient. The scale of the ground proximity ratio $h/b$ is plotted in logarithmic

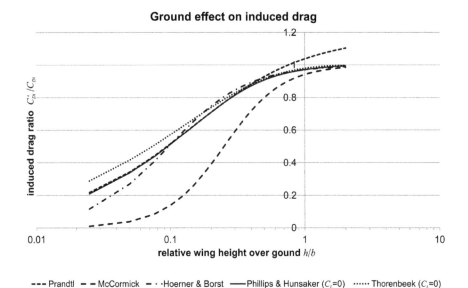

**Ground effect on induced drag**

--- Prandtl   – – McCormick   – · ·Hoerner & Borst   ——Phillips & Hunsaker ($C_L$=0)   ······ Thorenbeek ($C_L$=0)

**FIGURE 4.6** Comparison of different methods to calculate the effect of ground proximity on the drag coefficient.

scale to highlight the differences. The drawback of Prandtl's equation is that it is only valid below $2h/b \leq 0.5$ and it doesn't approach the value of 1 for higher altitudes. The other formulas except for McCormick's approximate Prandt's solution in low altitudes but asymptotically approach 1. In general, it is seen that the ground effect vanishes approximately at an altitude of one span over ground.

Regarding the influence of the ground proximity on the lift coefficient, Pistolesi [10] performed detailed analysis regarding the change in lift coefficient using vortex lattice methodology. The derived method for an infinite wing obtains the lift in ground effect

$$\frac{C_L'}{C_L} = k(1 - kr) \tag{4.24}$$

with the influence factor on circulation

$$k = \frac{1 + \left(\dfrac{c}{4h}\right)^2 \cos^2\alpha}{1 + \left(\dfrac{c}{4h}\right)\sin\alpha} \tag{4.25}$$

and on velocity

$$r = \frac{\left(\dfrac{c}{4h}\right)\sin\alpha}{1 + \left(\dfrac{c}{4h}\right)\sin\alpha} \tag{4.26}$$

assuming an ideal lift slope according to incompressible potential theory of

$$\frac{\partial C_L}{\partial \alpha} = 2\pi \tag{4.27}$$

including a dominant effect of the ratio of altitude to wing chord $h/c$. It is evident that the body has to produce lift to enforce a ground effect. It is therefore the slope of the lift coefficient that is affected by the ground effect. For the linear region of the lift curve[5] (assuming $\alpha = 0$ as point in the linear range)

$$C_L = C_{L0} + \frac{\partial C_L}{\partial \alpha} \alpha \tag{4.28}$$

the lift slope in ground effect is derived to

$$\frac{\partial C_L'}{\partial \alpha} = 2\pi \left( 1 + \left( \frac{c}{4h} \right)^2 \right). \tag{4.29}$$

Katz & Plotkin [11] make an analysis of the change in lift in 2D for a flat plate using lumb vortex analysis same as Pistolesi. They give for the circulation in ground effect the relation

$$\Gamma' = \pi U_\infty c \sin\alpha \left( \frac{1 - 2\left(\frac{c}{4h}\right)\sin\alpha + \left(\frac{c}{4h}\right)^2}{1 - \left(\frac{c}{4h}\right)\sin\alpha} \right). \tag{4.30}$$

As the circulation affects also the onflow velocity, the lift is related to the circulation by

$$L' = \rho U_\infty \Gamma' \left( 1 - \frac{\Gamma'}{4\pi U_\infty h} \right). \tag{4.31}$$

Using eq. (4.30) this leads to the sectional lift coefficient

$$c_L' = 2\pi \sin\alpha \left[ 1 - 2\frac{c}{4h}\sin\alpha + \left(\frac{c}{4h}\right)^2 (1 + \sin^2\alpha) + O\left(\left(\frac{c}{h}\right)^3\right) \right]. \tag{4.32}$$

Gross & Traub [12] also formulate a simplified theory that takes into account the induced angle of attack as well as the reduction of the onflow velocity induced by the mirrored lifting vortex. In contrast to Pistolesi [10] and Katz & Plotkin [11], they base their relation on the real lift slope gradient of the airfoil to improve the prediction

$$\frac{C_L'}{C_L} = \left[ 1 + \frac{1}{2\pi}\left(\frac{c}{4h}\right)\frac{\partial C_L}{\partial \alpha}\cos\left(\tan^{-1}\left(\frac{c}{4h}\right)\right)\sin\left(\tan^{-1}\left(\frac{c}{4h}\right)\right) \right] \cdot \left[ 1 - \frac{1}{2\pi}\left(\frac{c}{4h}\right)\frac{\partial C_L}{\partial \alpha}\alpha^2 \right].$$

$$\tag{4.33}$$

Similar to the relation for induced drag, Phillips & Hunsaker [8] give an approximation based on the aspect ratio and the height related to wing span

$$\frac{C'_L}{C_L} = 1 + \frac{288\left(\frac{h}{b}\right)^{0.787} e^{\left[-9.14\left(\frac{h}{b}\right)^{0.327}\right]}}{\Lambda^{0.882}} \tag{4.34}$$

with an extension to account for tapered wings and high lift coefficients

$$\frac{C'_L}{C_L} = \left(1 + \delta_L \frac{288\left(\frac{h}{b}\right)^{0.787} e^{\left[-9.14\left(\frac{h}{b}\right)^{0.327}\right]}}{\Lambda^{0.882}}\right) \Big/ \beta_L$$

$$\delta_L = 1 - 2.25\left(\lambda^{0.00273} - 0.997\right)\left(\Lambda^{0.717} + 13.6\right) \tag{4.35}$$

$$\beta_L = 1 + \frac{0.269 C_L^{1.45}}{\Lambda^{3.18}\left(\frac{h}{b}\right)^{1.12}}.$$

Nevertheless, for this approximation, the ground effect vanishes for infinite aspect ratio.

Although the above formulae suggest an increase of lift coefficient due to a steeper lift slope, "wind-tunnel test on aircraft configurations with powerful high-lift systems suggest that the ground effect may be unfavourable" – (Foster & East [14] p. 3). Recant [13] measured the ground effect on the maximum lift coefficient on a wing with a NACA 23012 airfoil with two different types of flaps. Figure 4.7 reproduces the measured maximum lift coefficients depending on the ground proximity for the plain airfoil, the airfoil equipped with a split flap and a slotted flap. While for the plain wing the maximum lift coefficient is independent of the ground proximity, flapped wings drop in maximum lift coefficient. Here, tapered wings show a reduced sensitivity on the ground effect.

Foster and East [14] analyze the ground effect of a finite span wing at high lift coefficients. They account for the different effects of the bound lifting vortex and the trailing vortices. Assuming the dependency of the lift slope on the aspect ratio by

$$\frac{\partial C_L}{\partial \alpha} = \frac{2\pi\Lambda}{\Lambda + 2}, \tag{4.36}$$

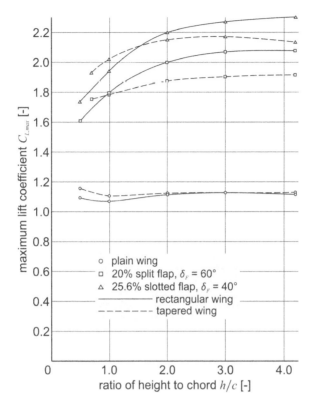

**FIGURE 4.7** Effect of ground proximity on maximum lift coefficients of NACA 23012 wings. (From Recant [13], figure 10.)

their analysis results in the relation

$$
C'_L = C_L \left[ 1 - \underbrace{\frac{\Lambda^2}{(\Lambda+2)\left(\Lambda^2+\left(\dfrac{4h}{c}\right)^2\right)}}_{\text{trailing vortices}} + \underbrace{\frac{\Lambda C_L}{\pi\left(\dfrac{4h}{c}\right)\sqrt{\Lambda^2+\left(\dfrac{4h}{c}\right)^2}}}_{\text{bound vortex}} \right]^{-1}, \qquad (4.37)
$$

where the positive (lift increasing) effect is attributed to the trailing vortices, while the negative (lift reducing) effect originates from the mirror of the bound vortex. "The favourable effect of ground proximity on the lift coefficient due to the trailing vortices is proportional to $C_L$ and the adverse effect due to the bound vorticity is proportional to $C_L^2$" – (Foster & East [14] p. 7). The adverse effect from the bound vorticity results from the reduction of flow speed, similar to the conclusions leading to eq. (4.31). At infinite aspect ratio, eq. (4.37) predicts for $\Lambda \to \infty$ the two-dimensional, or sectional, lift coefficient[6]

$$c_L' = c_L \left[ 1 + \frac{c_L}{\pi\left(\dfrac{4h}{c}\right)} \right]^{-1}. \tag{4.38}$$

Figure 4.8(a) depicts the relative change of the lift coefficient due to ground proximity based on eq. (4.37). It shows that the positive effect on lift coefficient postulated by Pistolesi exists mainly for low and medium lift coefficients. No positive effect is seen for lift coefficients above $C_L \geq 2$. At higher lift coefficients, the loss of lift coefficients gets very significant for ground proximities less than $h/c \leq 1$.

Foster and East [14] additionally anticipate "that the principal effect of the ground proximity on the maximum lift coefficient will be to reduce it in proportion to the square of the effective streamwise velocity at the airfoil, since viscous effects are expected to limit the lift to a constant value of the lift coefficient based on the effective streamwise velocity" – (Foster & East [14] p. 10). As trailing vortices should be neglected in this case, they provide the relation

$$C_{L,\max}' = C_{L,\max} \left[ 1 + \frac{C_{L,\max}}{8\pi\left(h/c\right)} \right]^{-2}. \tag{4.39}$$

In Figure 4.8(b), the result of eq. (4.39) is shown. Resuming a maximum lift coefficient of a transport aircraft in the order of $C_{L,\max} = 2.5 \div 3.5$ at ground proximities less than $h/c \leq 4$, the drop in maximum lift coefficient exceeds 5%.

## 4.2 AERODYNAMIC PERFORMANCE INDICATORS

In the following, the flight phases described in Section 4.1.1 are analyzed regarding the influence factors of aerodynamics. The sequence is built up analogously to a flight mission starting from ground acceleration over climb and descent to landing.

### 4.2.1 TAKE-OFF

The complete take-off is the acceleration of the aircraft on ground up to achieving an initial climb state. It ends at the defined obstacle height of 35 ft. A good high-lift performance for this segment can be defined as short as possible runway length needed to take off. The take-off flight phase itself can be separated into several sub-segments, as depicted in Figure 4.9.

The first part is the ground roll with zero incidence where the aircraft is accelerated with full thrust until the nose can be inclined at the rotation speed $V_R$. The

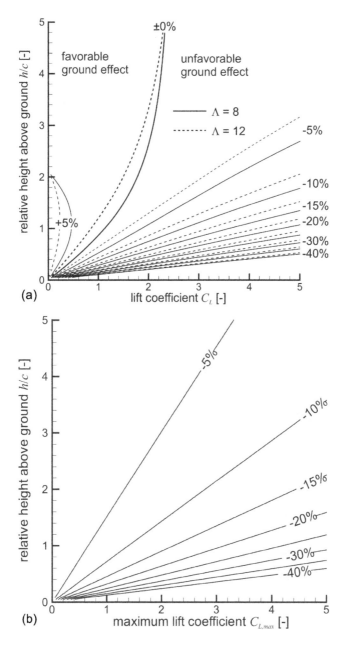

**FIGURE 4.8** Impact on ground proximity on lift coefficient and maximum lift coefficient of flapped airfoils according to eqs. (4.37) and (4.39). (Figure (a) extended diagram based on Foster & East [14], figure 2.)

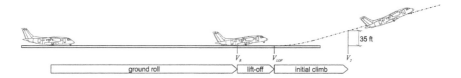

**FIGURE 4.9**  Segments of the take-off flight procedure.

runway length needed for this is depending on the acceleration and the end speed. As the acceleration can be expressed in the derivative of velocity versus length

$$a = \frac{dV}{dt} = V\frac{dV}{ds} = \frac{1}{2}\frac{d\left(V^2\right)}{ds}, \tag{4.40}$$

the runway length needed to accelerate the aircraft up to a certain speed can be calculated by the integration of the acceleration by

$$ds = \frac{1}{2a}d\left(V^2\right) \Rightarrow s = \frac{1}{2}\int_0^V \frac{1}{a}d\left(V^2\right). \tag{4.41}$$

The forces accelerating the aircraft on the ground are the sum of the thrust of the engines, the roll resistance of the wheels, and the aerodynamic drag force.

$$m\cdot a_{GR} = T - D - \mu_R\left(W - L\right). \tag{4.42}$$

The roll friction coefficient is between $\mu_R = 0.02$ for paved concrete runways, $\mu_R = 0.05 \div 0.1$ for grass landing fields, and $\mu_R = 0.3$ for soft grounds[7]. It is important to notice that the rolling resistance is reduced by the increasing lift force as the pressure on the wheels is released. Respecting that mass and weight are coupled by the gravity acceleration and respecting the dependency of the aerodynamic forces on the velocity, the ground acceleration is given by

$$a_{GR} = \left(\frac{T}{W} - \mu_R\right)g + \frac{\rho}{2}\frac{A_{ref}}{W}g\left(\mu_R C_{L0}' - C_D'\big|_{\alpha=0}\right)V^2 \tag{4.43)[8]}$$

with the aerodynamic conditions at zero incidence[9]. For an aircraft with jet engines, the thrust can be assumed constant, and eq. (4.41) can be solved analytically for the ground acceleration[10]

$$s_{GR} = \frac{1}{\rho g\dfrac{A_{ref}}{W}\left(\mu_R C_{L0}' - C_D'\big|_{\alpha=0}\right)}\cdot\ln\left(1 + \frac{\dfrac{\rho}{2}\dfrac{A_{ref}}{W}\left(\mu_R C_{L0}' - C_D'\big|_{\alpha=0}\right)}{\dfrac{T}{W} - \mu_R}V_R^2\right). \tag{4.44}$$

If the thrust is depending on the velocity as in propeller-driven aircraft, no analytical solution is possible.

The condition after the rotation depends on the aircraft configuration. Aircraft with long fuselages can be limited by the rotation incidence as the tail should not strike the ground. In this case, the incidence after rotation is defined by the aircraft geometry. In the other case, the incidence of the aircraft after rotation is related to the lift coefficient at lift-off resulting in a flight speed that may not be less than 10% above stall speed.

$$V_{LOF} \geq 1.1 V_{SR} \tag{4.45}$$

$$C_L'(\alpha_{LOF}) = \frac{TOW}{\frac{\rho}{2} A_{ref} V_{LOF}^2} < \min\left(C_{L,\max}'/1.1^2, C_L'(\alpha_{\max})\right). \tag{4.46}$$

The acceleration between rotation speed and lift-off speed is therefore defined by the aerodynamic state at the corresponding angle of attack

$$a_{LOF} = \left(\frac{T}{W} - \mu_R\right)g + \frac{\rho}{2}g\frac{A_{ref}}{W}\left(\mu_R C_L'(\alpha_{LOF}) - C_D'(\alpha_{LOF})\right)V^2 \tag{4.47}$$

with the analytical solution for constant thrust

$$
\begin{aligned}
S_{LOF} = {} & \frac{1}{\rho g \dfrac{A_{ref}}{W}\left(\mu_R C_L'(\alpha_{LOF}) - C_D'(\alpha_{LOF})\right)} \\
& \times \ln\left[\frac{\left(\dfrac{T}{W} - \mu_R\right) + \dfrac{\rho}{2}\dfrac{A_{ref}}{W}\left(\mu_R C_L'(\alpha_{LOF}) - C_D'(\alpha_{LOF})\right)V_{LOF}^2}{\left(\dfrac{T}{W} - \mu_R\right) + \dfrac{\rho}{2}\dfrac{SA_{ref}}{W}\left(\mu_R C_L'(\alpha_{LOF}) - C_D'(\alpha_{LOF})\right)V_R^2}\right].
\end{aligned}
\tag{4.48}
$$

Finally, the aircraft has to perform a short initial climb to an altitude 35 ft above any obstacle along the flight path. Within this phase, the aircraft gets out of ground effect with further acceleration up to the minimum climb speed, which is 13% above the stall speed.

$$V_2 \geq 1.13 V_S. \tag{4.49}$$

As the effect of the speed difference between lift-off and climb speed is negligible, the distance can be calculated over the ground effect only. According to eq. (4.5), the glide path angle in this sub-segment is given by

$$\sin\gamma = \frac{T}{W}\cos\alpha - \frac{\left(C_{D0} + \dfrac{C_L^2}{e\pi\Lambda'(h)}\right)}{C_L'(h)}. \tag{4.50}$$

The variation of ground distance with height is for small glide path angles

$$\frac{ds}{dh} = \frac{1}{\tan\gamma} \approx \frac{1}{\sin\gamma} \tag{4.51}$$

and the ground distance for the initial climb can be approximately integrated by

$$s_{IC} = \int_{0}^{h_{obstacle}+35\,ft} \frac{W}{T - \frac{\rho}{2} A_{ref} V_2^2 \left( C_{D0} + \frac{\left(C_L'(h)\right)^2}{e\pi\Lambda'(h)} \right)}\, dh. \tag{4.52}$$

The overall take-off field length (TOFL) is the sum of all three essential sub-segments

$$TOFL = s_{GR} + s_{LOF} + s_{IC}. \tag{4.53}$$

As the calculation of the take-off path gets rather complicated, in preliminary aircraft design, an approximation formula is used that establishes an averaged aerodynamic behavior at 70%[11] of the lift-off flight speed

$$TOFL = \frac{WV_{LOF}^2}{2gF_{eff}}; \quad F_{eff} = \left[T - D - \mu_R\left(W - L\right)\right]\big|_{V=0.7V_{LOF}}. \tag{4.54}$$

## 4.2.2   CLIMB

The climb performance of an aircraft can have two distinctive objectives. Either it is intended to obtain the steepest climb, which is correlating to maximizing the climb gradient. Or the intention is to achieve the cruising altitude as fast as possible, thus minimizing the time to climb. For the fastest climb, the climb rate is, therefore, to be maximized. The general relation regarding climb gradient and climb rate as presented in Section 4.1 by eq. (4.2) and eq. (4.6), respectively, apply. Especially for the climb performance, the simplification of low flight path angles is no longer valid and the balance between lift, drag, and thrust involves a significant contribution of the thrust to support lift.

Additionally, the available thrust or power is depending on the flight speed and engine type. It is therefore worth to compare the available thrust/power to the energy/power needed to climb the aircraft.

Derived from eq. (4.3), for a given speed, the weight of the aircraft relates to the lift coefficient by

$$W = \frac{1}{\cos\gamma}\left[C_L(\alpha)\cdot\frac{1}{2}\rho\cdot V^2\cdot A_{ref} + T(V)\cdot\sin(\alpha+\varepsilon_E)\right]. \tag{4.55}$$

This balance defines the angle of attack at which the aircraft is operating. At the same time, the thrust contribution in flight direction compensates the drag, given by

$$T(V)\cdot\cos(\alpha+\varepsilon_E) = C_D(\alpha)\cdot\frac{1}{2}\rho\cdot V^2\cdot A_{ref} + W\cdot\sin\gamma \tag{4.56}$$

and rearranged for the climb gradient

$$\gamma = \sin^{-1}\left|\frac{T(V)\cos(\alpha+\varepsilon_{T_E})-C_D(\alpha)\cdot\frac{1}{2}\rho\cdot V^2 \cdot A_{ref}}{W}\right|. \tag{4.57}$$

In this set of equations, the angle of attack has to be iterated taking into account the actual aerodynamic aircraft data. Nevertheless, the direct dependence of the climb gradient on the excess thrust is obvious. Therefore, the maximum climb gradient is achieved at the maximum excess thrust.

According to eq. (4.6), the climb rate is directly related to the flight speed and the climb gradient

$$w = V \cdot \sin\gamma \tag{4.58}$$

and, using eq. (4.57), this results in

$$w = \frac{T(V)\cdot V\cdot\cos(\alpha+\varepsilon_T)-C_D(\alpha)\cdot\frac{1}{2}\rho\cdot V^3 \cdot A_{ref}}{W}. \tag{4.59}$$

As the engine power is defined as

$$P = T\cdot V, \tag{4.60}$$

the maximum climb rate is therefore linked to maximum excess power.

In order to understand the optimum operating conditions for steepest climb or fastest climb, it is necessary to have a look on the evolution of drag depending on the flight speed. First, it has to be remembered that a low flight speed corresponds to a high lift coefficient. The major contributions to drag can be divided into the lift-induced drag and the parasitic drag. The parasitic drag relates to the portion of the drag not influenced by the lift and can therefore be estimated as

$$D_{\text{parasitic}} = C_{D0}\cdot\frac{\rho}{2}A_{ref}\cdot V^2. \tag{4.61}$$

The lift-induced drag coefficient depends on the lift coefficient, generally assumed to relate quadratically by

$$C_{Di} = \frac{C_L^2}{e\pi\Lambda}. \tag{4.62}$$

The lift coefficient – if not determined by eq. (4.55) – can be simplified related to the velocity by

$$C_L = \frac{2W}{\rho A_{ref}}\frac{1}{V^2} \tag{4.63}$$

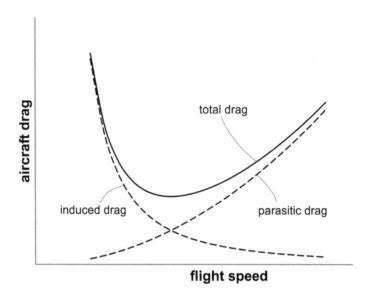

**FIGURE 4.10** Contribution of lift-induced drag and parasitic drag to the total aircraft drag depending on flight speed.

so that the lift-induced drag depends on velocity by

$$D_{\text{induced}} = \frac{1}{e\pi\Lambda} \frac{2W^2}{\rho A_{ref}} \frac{1}{V^2}. \tag{4.64}$$

Figure 4.10 sketches the principal dependency of these two major contributors to the total aircraft drag. As indicated by the equations above, the lift-induced drag reduces with increasing flight speed due to the reduction of the lift generating circulation. The parasitic drag increases with the square of the velocity. The graph of the total drag shows a distinct minimum of the drag that corresponds to the minimum requirement on propulsion force to enable a non-accelerated level flight.

For a turbojet-driven aircraft, the thrust available can be "approximated as being constant with airspeed and directly proportional to air density" – (Phillips [15] p. 244). Thus, the available power grows linearly with flight speed. Figure 4.11 shows the determination of the optimum speeds for steepest climb and fastest climb for such an aircraft. The best climb rate is obtained at the maximum excess power, the best climb gradient at the maximum excess thrust. Both velocities are higher than the speed for minimum required power. For turbojet aircraft, the speed for steepest climb is associated with the minimum thrust requirement.

For propeller-driven aircraft, the situation slightly differs. "An engine combined with a variable pitch propeller can be adjusted [...] to produce an available power that is nearly independent of airspeed over the range of airspeeds encountered in normal flight. [...] this type of power plant [can be approximated] as a device that produces an available power that is constant with airspeed" – (Phillips [15] p. 235). Figure 4.12 shows the corresponding determination of the characteristic speeds for climb. The

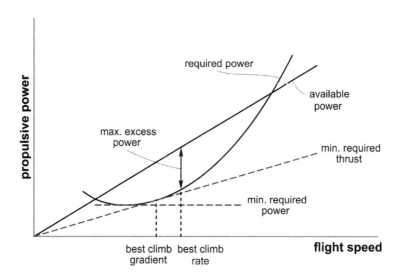

**FIGURE 4.11**  Determination of speeds for best climb gradient and best climb rate for a turbojet-driven aircraft.

flight speed for steepest climb is less than that for minimum required thrust. The speed for fastest climb corresponds to the speed of minimum required power. As real engine performance varies from the constant power approximation, the best climb speed will be at least near this condition.

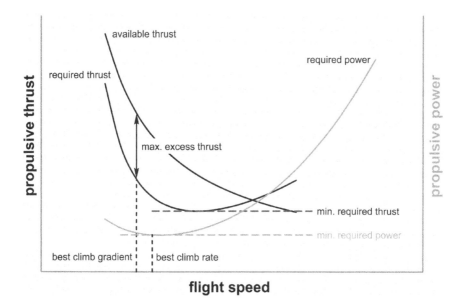

**FIGURE 4.12**  Determination of speeds for best climb gradient and best climb rate for a propeller-driven aircraft.

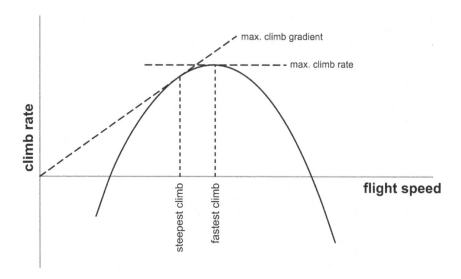

**FIGURE 4.13**  Velocity polar during climb at maximum power indicating the characteristic speeds for steepest and fastest climb.

Figure 4.13 depicts, in a general way, the dependence of the climb rate on the flight speed at a maximum power setting. The maximum climb rate is directly seen. According to eq. (4.58), the climb gradient at any speed is the angle of a line connecting the origin with the corresponding point on the curve. The maximum climb gradient is therefore indicated by the tangent to the climb rate curve through the origin. Again, it is clearly seen that the speed for steepest climb is always lower than for fastest climb.

An additional impact regarding high-lift systems is the choice of the best flap setting for climb. In general, a higher flap setting offers a lower take-off speed. But as such a higher flap setting usually induces a higher required thrust due to a higher drag coefficient, in general, the climb gradient or climb rate will be less than for a lower flap setting. In the opposite way, a lower flap setting results in a higher take-off speed and a better climb performance. Figure 4.14 illustrates the dependency of the overall take-off path depending on the flap setting. In general, a lower flap setting leads to a longer ground run and a steeper climb than a higher flap setting. The

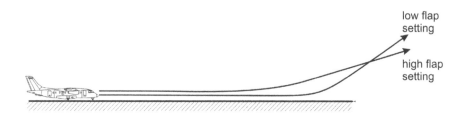

**FIGURE 4.14**  Impact of flap setting on climb performance.

cross-over of both flight paths is somewhere along the climb flight path. The optimal choice of a flap setting strongly depends on the actual aircraft configuration, runway requirements, and environmental conditions such as altitude and temperature. For short runways, it can be necessary to select a higher flap setting. A similar decision may be beneficial with respect to noise certification if the cross-over of the flight paths is at a higher altitude.

### 4.2.3 Approach

The approach of an aircraft to an airport is strongly prescribed by the local air traffic authority. The aircraft has to follow the required actions to be scheduled into the procedures of multiple aircraft of very different types. Aircraft are therefore assigned into approach categories based on their minimum approach speed that is

$$V_{REF} = 1.23 V_{SR}. \qquad (4.65)$$

The International Civil Aviation Organization (ICAO) defines five categories according to Table 4.2. These categories allow the air traffic control to plan for the assigned air space and the scheduling of aircraft for landing.

The current standard procedures for approaches of civil transport aircraft are given in Figure 4.15. The procedures are characterized by the point of descent when the engine thrust is reduced. At the deceleration point, the flight speed is reduced. Then, during the deceleration, the high-lift settings are utilized to further reduce the flight speed. At an altitude of 3000 ft, the aircraft is brought on the final approach path given by the instrument landing system (ILS). The standard ILS glide path

---

**TABLE 4.2**

**Aircraft Approach Classes Defined in the ICAO Aircraft Operations (PANS-OPS) [16]**

| Aircraft Category | $V_{REF}$ | Range of Speeds for Initial Approach | Range of Final Approach Speeds | Maximum Speeds for Circling | Maximum Speeds for Intermediate Missed Approach | Maximum Speeds for Final Missed Approach | Typical Aircraft in This Category |
|---|---|---|---|---|---|---|---|
| A | <91 | 90–150 | 70–110 | 100 | 100 | 110 | Small Single Engine |
| B | 91–120 | 120–180 | 85–130 | 135 | 130 | 150 | Small Multi Engine |
| C | 121–140 | 160–240 | 115–160 | 180 | 160 | 240 | Airline Jet |
| D | 141–165 | 185–250 | 130–185 | 205 | 185 | 265 | Large/ Military Jet |
| E | 166–210 | 185–250 | 155–230 | 240 | 230 | 275 | Special Military |

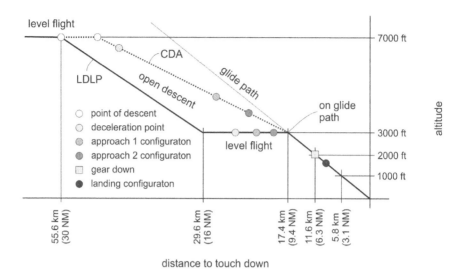

**FIGURE 4.15** Standard approach procedures. (Diagram according to König [18].)

angle is $\gamma = 3°$. Some airports (e.g., Frankfurt/Main) have already introduced a slightly higher inclination of the glide path at $\gamma = 3.2°$ to reduce the noise perception on the airport surrounding. The airport London City (EGLC) requires the steepest descent glide path at $\gamma = 5.5°$ [17].

The most common procedure is the so-called low-drag/low-power approach (LDLP). It includes a very late deceleration and deflection of the high-lift devices. By this, the approach is very efficient in terms of traveling time. The so-called continuous descent approach (CDA) is another utilized approach where the deceleration is more spread along the descent and the high-lift systems are deflected earlier. In comparison to the LDLP, this approach takes a longer time, but less fuel consumption as the intermediate level flight is avoided. Additionally, due to the higher altitudes and the lower thrust setting, this approach is used to reduce the noise perception at a distance of 10–30 nautical miles from the airport.

## 4.2.4 LANDING

The design criterion associated with landing is the runway length needed. The flight path of the landing after passing the decision altitude of 50 ft above ground can be subdivided into four parts, illustrated in Figure 4.16. The first part is a steady gliding flight until the aircraft has to be leveled out. The transition is usually approximated by an arc at constant flight speed, followed by the flare out where the aircraft speed reduces from the approach speed to the touch-down speed. After touch-down, the aircraft is actively retarded by using brakes, spoilers, and reverse thrust until standstill. The distances can be calculated using similar formulae as for the take-off.

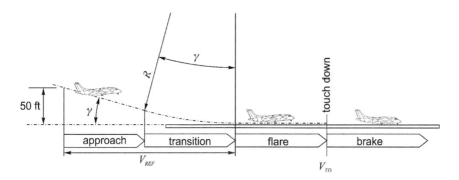

**FIGURE 4.16**   Segments of the flight path during landing.

For the distance before touch-down, one significant parameter is the height for initiating the leveling out, which is determined by the load factor and thereby the radius of the flight path.

$$R = \frac{V_{REF}^2}{g(n_z - 1)}. \tag{4.66}$$

The load factor for the transition into flare is usually assumed[12] to be $n_z = 1.2$. The corresponding height above ground to initiate flare is geometrically given by

$$h_f = R(1 - \cos\gamma) \tag{4.67}$$

depending on the flight path angle. The distance from passing the 50 ft altitude until the initiation of flare is therefore

$$s_A = \frac{50\,\text{ft} - h_f}{\tan\gamma} \tag{4.68}$$

and the distance of the leveling out

$$s_R = R\sin\gamma. \tag{4.69}$$

These distances are independent of the flight speed.

After leveling out, the aircraft is decelerated from approach speed to touch-down speed by aerodynamic drag only. The corresponding deceleration is

$$a_{FL} = \left(\frac{T}{W}\right)g + \left[\frac{\rho}{2}\frac{A_{ref}}{W}gC_D'\right]V^2. \tag{4.70}$$

The touch-down speed should not be too low and for transport aircraft a safety margin of 15% above stall speed is recommended – (see Jenkinson et al. [2] p. 242, Phillips [15] p. 313). Nevertheless, as for the take-off, the maximum achievable angle of attack may be an additional limit to prevent tail strike. The corresponding lift coefficient is therefore

$$C'_{L,TD} = \min\left(C'_{L,\max}\Big/1.15^2\,;\,C'_L\left(\alpha_{\max}\right)\right). \tag{4.71}$$

An approximate formula is obtained by assuming the aerodynamics not to vary much and therefore a constant deceleration.

$$s_{FL} \approx \frac{1}{2g}\frac{W}{T-D}\left(V_{\mathrm{Ref}}^2 - V_{TD}^2\right) \tag{4.72}$$

After touch-down, the aircraft is in a short free roll. Regulations foresee to account for 2 seconds the pilot needs to act for further deceleration by brakes, spoilers, and reverse thrust. The corresponding ground roll is commonly assumed to be non-decelerated and the corresponding distance is given by

$$s_{FR} = t\cdot V_{TD} = t\sqrt{\frac{2}{\rho}\frac{W}{S}\frac{1}{C'_{L,TD}}}. \tag{4.73}$$

The remaining distance using all deceleration devices is then with the deceleration

$$a_b = \left(\frac{T}{W} - \mu_{R,b}\right)g + \frac{\rho}{2}\frac{A_{ref}}{W}g\left(\mu_{R,b}C'_{L0,SP} - C'_{D,SP}\big|_{\alpha=0}\right)V^2. \tag{4.74}$$

The friction coefficient using all brakes is depending on the runway condition[13]. While on a dry runway, the equivalent friction coefficient can achieve values of $\mu_{R,b} = 0.5$, this is naturally lower for wet ($\mu_{R,b} = 0.3$) or even icy runway conditions ($\mu_{R,b} = 0.1$). The aerodynamic coefficients have to account for the additional drag of the spoilers and their effect on lift (lift dump). The thrust is either the remaining idle thrust of the engine or the (negative) reverse thrust[14]. The potential of reverse thrust is depending on the engine type. Typical values are listed in Table 4.3.

## 4.2.5 SHORT TAKE-OFF AND LANDING

One of the first definitions of Short Take-Off and Landing (STOL) has been given by John H. Shaffer, US Federal Aviation Administration (FAA) [20]: "an airplane that

**TABLE 4.3**
**Reverse Thrust Potential Depending on Engine Type[15]**

| Engine Type | Reverse Thrust Potential [% max. thrust] |
| --- | --- |
| Propeller | 40 |
| Turbojet | 40–50 |
| Turbofan | 60 |

can really get off from and land on a 2,500-foot strip and do it safely and efficiently" – (Shaffer [20] p. 55). Later, this FAA position was reformulated by John Kern in the scope of a hearing in front of the US Congress House of Representatives, Committee on Science and Technology [21]: "A STOL aircraft is an aircraft with a certified performance capability to execute approaches along a glideslope of 6 degrees or steeper and to execute missed approaches at a climb gradient sufficient to clear a 15:1 missed approach surface at sea level" – (Kern [21] p. 182). In conclusion, this definition given by the FAA in 1984 relates more to what is today called Steep Approach Landing (SAL) capability.

More descriptive is a definition on airport site formulated by the International Civil Aviation Organization (ICAO) [22]: "The STOLport design aeroplane is assumed to be an aeroplane that has a reference field length of 800 m or less" – (ICAO STOLport manual [22] p. 1). In this context, it must be understood that the reference field length is defined to be the actual required landing distance divided by 0.6. Thus, the actual field length required for a safe take-off or landing has to be less than 480 m. The threshold of 800 m results from the definition of an airport, to have at least one runway with more than 800 m length equipped for instrument landing approach. The STOLport definition enables commercial air transport with STOL capable aircraft on smaller airfields.

Another definition of STOL operation is given in the US Department of Defense Dictionary [23]. There STOL is defined as the ability of "an aircraft to clear a 50-foot (15 meters) obstacle within 1,500 feet (450 meters) of commencing takeoff or in landing, to stop within 1,500 feet (450 meters) after passing over a 50-foot (15 meters) obstacle" – (US DoD Dictionary [23] p. 217). This definition is close, but not exactly the same as the ICAO definition.

In Europe, a definition of STOL is neither available in the Certification Specifications (CS-25) nor in the aircraft operation requirements (Part-OPS). Nevertheless, similar to the FAA, the CS-25 Appendix Q defines what is called a Steep Approach Landing (SAL) capability [24]: "aeroplane to obtain approval for a steep approach landing capability using an approach path angle greater than or equal to 4.5° (a gradient of 7.9%)" – (EASA CS-25 Appendix Q [24] p. 1). The European definition is thereby less stringent than the FAA's understanding.

At least there is no unique or common definition of STOL in terms of certification specifications or airworthiness instructions, neither in Europe nor in Northern America. But, a large number of technological inventions are labeled with STOL capabilities. In conclusion, in most cases this label highlights "an airplane with field length requirements substantially less than those of a CTOL [Conventional Take-Off and Landing] airplane, of the same payload, range, and speed" – (Sincoff & Dajani [25] p. 17).

## NOTES

1. Wieselsberger [3] and Prandtl [4] use the distance of the two circulation centers in their original formulations, so twice the distance of the wing to the ground. As most other formulations use the height of the wing above ground, this has been adopted here.

2. This equation is eq. (7.4) on page 360 of the 2nd edition. Philipps & Hunsaker [8] state the equation in the 1st edition being $\Lambda' = \Lambda\left[1+\left(16\dfrac{h}{b}\right)^2\right]\Big/\left(16\dfrac{h}{b}\right)^2$ at the 1st print and $\Lambda' = \Lambda\left[1+\left(16\dfrac{h}{\pi b}\right)^2\right]\Big/\left(16\dfrac{h}{\pi b}\right)^2$ past the 18th print. McCormick's [5] solution is derived "simply by determining the effect of the ground on the downwash midspan between a pair of vortices representing a completely rolled-up vortex system. Image vortices similar to [...] [Figure 4.5] were used to account for the ground". The method is similar to the solution of Suh & Ostowari [6].

3. The equation cited at Phillips & Hunsacker [8] as eq. (7) uses the ratio of span $b$ and aspect ratio $\Lambda$ instead of the geometric mean chord $c$ in the lift dependent term.

4. Hoerner [9], eq. (8), page 20–10. Phillips & Hunsaker [8] cite this equation from the 1st ed. (1975) as $\dfrac{\left(C_{D,i}/C_L^2\right)'}{C_{D,i}/C_L^2} = \dfrac{33\left(\dfrac{h}{b}\right)^{3/2}}{33\left(\dfrac{h}{b}\right)^{3/2}+1}$ assuming a linear dependency of the induced drag on the induced incidence.

5. As pointed out by Katz & Plotkin [11] p. 131, the exact potential solution for the lift coefficient is $C_L = 2\pi\sin\alpha$. Eqs. (4.27) and (4.28) therefore assume small angles of attack where the approximation $\alpha \approx \sin\alpha$ is valid, which is usually denoted as the linear lift regime.

6. Foster & East [14] refer to the exact solution for the single vortex section lift coefficient at infinite aspect ratio of $c_L' = c_L\left[1+\dfrac{c_L}{2\pi\left(\dfrac{4h}{c}\right)}\right]^{-2}$ mentioning that the provided formulation is a close approximation in the area of interest.

7. Values from Jenkinson et al. [2] (p. 230). Phillips [15] states $\mu_R = 0.04$ for paved surfaces and $\mu_R = 0.3$ for grass (p. 302).

8. Theoretically, there is the case that the velocity-dependent effect on the acceleration vanishes. This is exactly the case for $^{C_L}\!/_{C_D} = 1\!/_{\mu_R}$. This would mean a lift-to-drag ratio of 50 for paved runways but only $10 \div 20$ for grass fields. It has only to be taken into account as it forms a singularity in eq. (4.44) but then even the integration is simplified.

9. Note that while the notation "zero lift coefficient" $C_{L0}$ is for the lift coefficient at zero angle of attack, the notation "zero drag coefficient" $C_{D0}$ is used for the drag coefficient at zero lift coefficient. The latter is used to identify the lift-induced drag coefficient.

10. Formula according to Jenkinson et al. [2], eq. (10.5), p. 230, but there, the division sign in the aerodynamic factor eq. (10.4) is missing.

11. Value from Phillips [15] p. 307, McCormick [5] states a value of $1/\sqrt{2}$ (p. 364), Jenkinson et al. [2] state a value of 0.707 (p. 229).

12. Value from Jenkinson et al. [2].

13. Value from Jenkinson et al. [2] p. 243, Phillips [15] uses a value of $\mu_{R,b} = 0.4$.

14. The FAR requires that, for estimating the needed runway length, a thrust reverser must not be accounted for, as an engine failure might prevent its usage.

15. data according to Anderson [19] p. 370.

## REFERENCES

[1] Meredith PT (1993) Viscous Phenomena Affecting High-Lift Systems and Suggestions for Future CFD Development, no. 19 in High-Lift System Aerodynamics, AGARD CP 515.

[2] Jenkinson L, Simpkin P, Rhodes D (1999) Civil Jet Aircraft Design, Butterworth-Heinemann, London, ISBN 0-340374152-X.

[3] Wieselsberger C (1922) Wing Resistance Near the Ground, NACA-TM-77, translation from Zeitschrift für Flugtechnik und Motorluftschiffahrt 10(1921), pp. 145–147.

[4] Prandtl L (1924) Induced Drag of Multiplanes, NACA TN 182, translation from Technische Berichte herausgegeben von der Flugzeugmeisterei der Inspektion der Fliegertruppen, Charlottenburg. Bd. III (1918) pp. 309–315.

[5] McCormick BW (1995) Aerodynamics, Aeronautics, and Flight Mechanics, 2nd ed., John Wiley & Sons, New York.

[6] Suh YB, Ostowari C (1988) Drag Reduction Factor due to Ground Effect, Journal of Aircraft 25(11), pp. 1071–1072.

[7] Torenbeek E (1982) "Ground Effects," Synthesis of Subsonic Airplane Design, Delft Univ. Press, Delft, The Netherlands, pp. 551–554.

[8] Phillips WF, Hunsaker DF (2013) Lifting-Line Predictions for Induced Drag and Lift in Ground Effect, Journal of Aircraft 50(4), pp. 1226–1233.

[9] Hoerner SF, Borst HV (1985) Fluid Dynamic Lift, 2nd ed., Hoerner Fluid Dynamics, Bakersfield, CA.

[10] Pistolesi E (1937) Ground Effect – Theory and Practice, NACA-TM-828, translation from Pubblicazioni della R. Scuola d'Ingegneria di Pisa series VI(261), pp. 1–25.

[11] Katz J, Plotkin A (2001) Low Speed Aerodynamics, 2nd ed., Cambridge University Press, Cambridge, UK.

[12] Gross J, Traub LW (2012) Experimental and Theoretical Investigation of Ground Effect at Low Reynolds Numbers, Journal of Aircraft 49(2), pp. 576–586.

[13] Recant IG (1939) Wind-Tunnel Investigation of Ground Effect in Wings with Flaps, NACA TN 705.

[14] Foster DN, East LF (1976) The Theoretical Effect of Ground Proximity on a High-Lift Wing, RAE TR 76139.

[15] Phillips WF (2004) Mechanics of Flight, John Wiley & Sons, Hoboken, NJ, ISBN 0-471-33458-8.

[16] ICAO (2006) Aircraft operations, PANS-OPS, ICAO Doc 8168.

[17] Civil Aviation Authorities (2016) EGLC AD 2.20 Local Traffic Regulations, United Kingdom AIP – EGLC – London City, p. 8.

[18] König R (2007) Umsetzung von Flugverfahren/Flight Management Systeme (Teil 1), Abschlussveranstaltung "Leiser Flugverkehr II," Göttingen.

[19] Anderson JD (1999) Aircraft Performance and Design, McGraw-Hill, Boston, MA, ISBN 0-07116010-8.

[20] Shaffer JH (1969) Role of General Aviation in the National Transportation System, In: The Role of General Aviation, Hearing before the Subcommittee on Aviation of the Committee on Commerce, US House of Representatives, Ninety-first Congress, second session, pp. 53–65.

[21] Kern J (1984) Aircraft Navigation and Landing Technology: Status of Implementation Hearing before the Subcommittee on Transportation, Aviation, and Materials of the Committee on Science and Technology, US House of Representatives, Ninety-eighth Congress, second session, July 24, 1984, pp. 181–191.

[22] ICAO (1991) STOLport Manual, 2nd edition, ICAO Doc. 9150-AN/899.

[23] Department of Defense (2010) Dictionary of Military and Associated Terms, Joint Publication 1-02.

[24] European Aviation Safety Agency (2015) Certification Specifications and Acceptable Means of Compliance for Large Aeroplanes, CS-25 Amendment 17, Appendix Q.

[25] Sincoff MZ, Dajani JS (1975) General Aviation and Community Development, NASA contract report NGT 47-003-028.

# 5 Passive High-Lift Systems

## NOMENCLATURE

| | | | | | |
|---|---|---|---|---|---|
| $c$ | – | Airfoil chord | $\mathbf{u}$ | $m/s$ | Local velocity vector |
| $c_F$ | – | Flap chord | $x, y, z$ | $m$ | Coordinate |
| $C_L$ | – | Lift coefficient | $\alpha$ | ° | Angle of attack |
| $C_{L,\max}$ | – | Maximum lift coefficient | $\beta$ | ° | Skew angle |
| $c_N$ | – | Normal force coefficient | $\Gamma$ | $m^2/s$ | Circulation |
| $c_p$ | – | Pressure coefficient | $\delta$ | ° | Deflection angle |
| $\bar{c}_p$ | – | Canonical pressure coefficient | $\delta$ | $m$ | Boundary layer thickness |
| $c_p^*$ | – | Critical pressure coefficient | $\mu$ | $kg/m\ s$ | Viscosity |
| $d$ | $m$ | Distance | $\rho$ | $kg/m^3$ | Density |
| $D$ | $m$ | Separation distance | $\tau_w$ | $N/m^2$ | Wall shear stress |
| $g$ | $m$ | Gap size | $\omega$ | $1/s$ | Rotation |
| $h$ | $m$ | Height | $\nabla$ | | Gradient operator |
| $l$ | $m$ | Length | $\Delta$ | | Difference operator |
| $M$ | – | Mach number | $\delta$ | | Boundary operator |
| $\mathbf{n}$ | $m^2$ | Surface normal vector | $\times$ | | Vector cross product |
| $p$ | | Static pressure | $_{DN}$ | | Droop nose |
| $r$ | $N/m^2$ | Radius | $_F$ | | Flap |
| $R$ | $m$ | Vortex core radius | $_{min}$ | | Minimum value |
| $Re$ | – | Reynolds number | $_{max}$ | | Maximum value |
| $\mathbf{s}$ | $m$ | Surface arc vector | $_s$ | | Slat |
| $S$ | $m^2$ | Surface | $_w$ | | Wing |
| $U$ | $m/s$ | Outer flow velocity | $_\infty$ | | Free stream |
| $u$ | $m/s$ | Local velocity | | | |

This chapter presents the passive means to increase the lift coefficient. Passive in this sense means that the change of the lift coefficient behavior is solely produced by changing the aerodynamic shape[1]. Passive systems do not alter the energy content of the flow in contrast to active high-lift systems, which are discussed in Chapter 6. In the following, three different classes of passive high-lift systems are discussed. First, it is distinguished between systems that open a gap and those that do not. In this context, the specific effects of the flow through gapped airfoils are discussed. The third class combines geometric features that create vortices that stabilize the flow to achieve higher lift coefficients.

Before looking at different high-lift devices, it is necessary to understand the principal difference of the effect of a high-lift system depending on where it is applied.

DOI: 10.1201/9781003220459-5

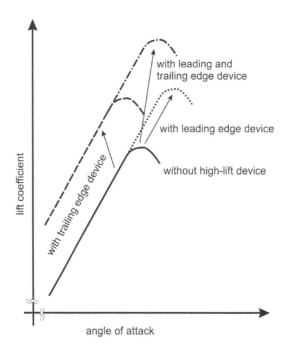

**FIGURE 5.1**    Principal effect of high-lift devices visualized by slat and flap effect on a high-lift wing.

Figure 5.1 shows the change of the lift coefficient versus the angle of attack for a wing with the high-lift system not deployed, deployed separately, and deployed together.

Basically, it can be seen that a leading-edge device, at first instance, elongates the linear range of the lift curve by extending the curve to higher angles of attack and, by this, generating a higher maximum lift coefficient. In the linear range of angles of attack of the clean wing without deployed high-lift devices, the leading-edge defect has almost no influence on the lift coefficient.

A trailing edge device alone creates an almost constant increase of lift coefficient over the complete range of angles of attack. The angle of attack for stall onset is thereby slightly reduced compared to the clean wing. The reason for the direct impact on lift coefficient at constant angle of attack results from the change of the Kutta condition due to the moved trailing edge. Since the reference axis for the angle of attack is retained from the clean airfoil, the high-lift system deflection at the trailing edge imposes an increased camber and an increased effective incidence.

The lift coefficient of an airfoil with a trailing edge device can be approximated in the linear range by the impact of the angle of attack and the flap deflection

$$C_L = \frac{\partial C_L}{\partial \alpha}\alpha + \frac{\partial C_L}{\partial \delta_F}\delta_F. \tag{5.1}$$

An approximation for the effectiveness of a trailing edge device can thereby be obtained by introducing an expression for the flap effectiveness

$$C_L = \frac{\partial C_L}{\partial \alpha}\alpha + \frac{\partial C_L}{\partial \alpha}\frac{\partial \alpha}{\partial \delta_F}\delta_F = \frac{\partial C_L}{\partial \alpha}\left(\alpha + \frac{\partial \alpha}{\partial \delta_F}\delta_F\right). \tag{5.2}$$

Glauert [1] gives an approximation of the flap effectiveness based on the size of the flap

$$\frac{\partial \alpha}{\partial \delta_F} = \frac{2}{\pi}\left(\sqrt{c_F/c\cdot(1-c_F/c)}+\sin^{-1}\left(\sqrt{c_F/c}\right)\right). \tag{5.3}$$

Figure 5.2 shows a comparison for the approximation with some early measurements. The general tendency is reproduced but the effectiveness is significantly overpredicted as expected for linear theory based on potential flow.

Deflecting high-lift devices both at the leading and trailing edge shows that the effects in principle superimpose. This does not necessarily indicate that the deflection of both types of systems is independent of each other. It may be that the deflection of a leading-edge device reduces the effect of a trailing edge device

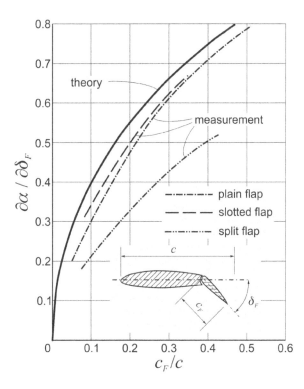

**FIGURE 5.2** Approximative flap effectiveness according to Glauert [1] in comparison to measurements. (According to Schlichting & Truckenbrodt [2], figure 12.8(a).)

or vice versa if not properly placed. But in case of proper placement, detrimental effects can be avoided.

## 5.1  SLOTLESS DEVICES

The first class of passive high-lift devices changes the airfoil shape without forming a slot or gap between the main airfoil and the device.

### 5.1.1  PLAIN FLAP

The plain flap – or camber flap – (Figure 5.3) is a device where the rear part of the airfoil is mounted on a hinge and deflectable at positive (downward) or negative (upward) deflection angles $\delta_F$. Its main aerodynamic principle is a simple variation of the camber of the airfoil. In fact, it is the oldest high-lift device. The first systematic investigation has been reported by Nayler, Stedman, and Stern in 1914 [3]. Nevertheless, Weyl [4] reports that already "the Le Blon monoplane built by Humber Ltd., and exhibited at the London Olympia Show of March 1910, actually had variable camber wings formed by an adjustable part of the trailing edge of the wing." – (Weyl [4] p. 294).

The impact of increasing the camber by a downward deflection of the rear portion of the airfoil shows the general trend of a trailing edge device.

Figure 5.4 shows measurements of the lift coefficient versus the angle of attack of a $c_F/c = 25\%$ plain flap at a NACA 0009 airfoil performed by Spearman [5]. At flap deflections, less than $\delta_F = 20°$, the lift coefficient in the linear range of the lift characteristics shows a linear dependency with the flap deflection angle and a slight reduction of the angle of attack where the maximum lift coefficient occurs.

The increase in lift coefficient at the same angle of attack can directly be attributed to the increase of circulation forced by the Kutta condition. Related to the unchanged flow direction, the flow must be turned more to keep the rear stagnation point at the downward moved trailing edge.

At higher deflection angles, the lift coefficient stagnates and no further increase is observed. This stagnation of the lift generation is attributed to the separation of the flow on the upper side of the plain flap. This separation develops immediately at the flap hinge and covers the whole flap's upper surface. In contrast to separations on a smooth surface, the separation is geometrically defined, limited extent, and therefore stable against changes of the angle of attack. The remaining airfoil stays attached and – together with the lift contribution of the pressure side – the lift coefficient stays

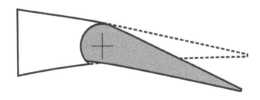

**FIGURE 5.3**  Principle of a plain flap.

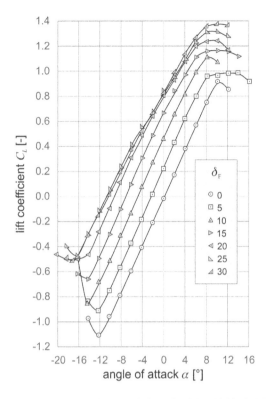

**FIGURE 5.4**    Aerodynamic section characteristics of NACA 0009 airfoil with a 25% plain flap. (From Spearman [5], figure 3.)

stable. Of course, the onset of the flap separation causes an increase in drag coefficient and pitch-up characteristics of the pitching moment.

Figure 5.5 shows pressure distributions at the same lift coefficient for different flap deflection angles of a $c_F/c = 20\%$ flap at a NACA 23012 airfoil, measured by Wenzinger & Rogallo [6]. Depending on the radius of the airfoil contour at the flap hinge, the local curvature induces a local suction peak at the flap hinge with a very limited extent in chord direction. This suction peak implies two additional effects. Downstream the hinge, the pressure gradient is now much steeper than for the undeflected flap. Therefore, the separation tends to start immediately after the hinge due to the strong pressure gradient. The flap, therefore, does not separate gradually as a usual trailing edge separation but very sudden like a leading-edge separation, but at the beginning of the flap. Upstream the hinge, the flow gets accelerated. Therefore, a separation of the upper side, starting from the trailing edge of the fixed airfoil, is prevented. The separation on the wing, therefore, starts not further downstream than the beginning of the reacceleration.

Fiecke [7] presented the potential of maximum lift coefficient increase of plain flaps of different sizes and different deflecting angles. There are several distinct properties seen in the corresponding diagram, reproduced in Figure 5.6. At low deflection angles below $\delta_F = 30°$, there is a linear dependency of maximum lift

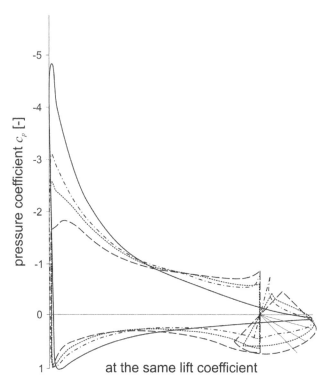

**FIGURE 5.5** Comparison of the pressure distribution on NACA 23012 airfoil with a 20% plain flap with that on the plain wing at same lift coefficient. (According to Wenzinger & Rogallo [6], figure 10 (a).)

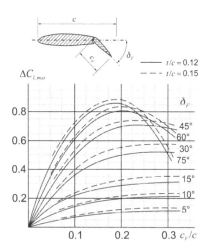

**FIGURE 5.6** Increase of maximum lift coefficient due to plain flap deflection depending on the relative flap length at different deflection angles for two airfoils of varying thickness. (Reproduced from Fiecke [7], figure 118.)

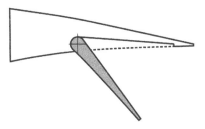

**FIGURE 5.7** Principle of a split flap.

coefficient increase with deflection angle. At relative flap chords above $c_F/c = 25\%$, the potential of lift generation levels out, and a larger flap is not able to produce increased maximum lift coefficient. At higher deflection angles, the linear growth of effectivity of the plain flap in terms of $\partial C_{L,max}/\partial\delta_F$ reduces as already seen before. At large flap chords, even a reduction of the maximum lift coefficient benefit is seen. The reason for the reduction of the lift generation capability is the separation of the flow on the upper side of the flap.

## 5.1.2 SPLIT FLAP

The split flap (Figure 5.7) has been developed from the camber tab by deflecting only the lower surface. The device was intended as not only a lift generating device but also as a speed brake at the same time, as the device generates a severe amount of additional drag. According to Weyl [4], the development of the split flap as a device to prevent premature stall was based on the discovery that an asymmetric thickening of the boundary layer – especially a thicker boundary layer on the lower side – causes the generation of lift. According to Irving [8], split flaps were first used somewhere on the lower side as a simple obstacle to thicken the boundary layer. The first patent is recorded for Royer [9], while Wright and Jacobs filed the first patent of the device that is today associated with the split flap in 1921 [10]. The diagram in their patent already shows the general aerodynamic principle, reproduced in Figure 5.8. They already discovered that, past the split flap, a recirculation area forms and therefore the low-pressure region on the upper side is extended past the wing trailing edge.

In 1932, first measurements were published by Schrenk and Gruschwitz [11] for an airfoil with a $c_F/c = 20\%$ split flap. Figure 5.9 shows the typical split flap behavior of the lift characteristic with respect to the angle of attack. Compared to the plain flap, the stall with deflected split flap changes its characteristic towards a more sudden drop in lift coefficient. The maximum lift coefficient increase is seen for a split flap deflected at $\delta_F = 90°^2$. The saturation of the lift increase potential is already given for a deflection of $\delta_F = 60°$.

Figure 5.10 shows effect of the split flap on the pressure distribution in terms of the spreading of the trailing edge pressure. This is the consequence of the existence of two trailing edges and the revocation of the Kutta condition. The large separation past the airfoil, due to the split flap, produces a large displacement of the flow and the joining of the mean flow that is usually located at the trailing edge appears much further

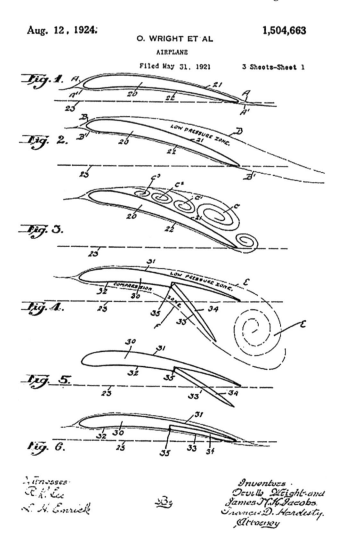

Aug. 12, 1924.                                                    1,504,663
          O. WRIGHT ET AL
              AIRPLANE
          Filed May 31, 1921        3 Sheets-Sheet 1

**FIGURE 5.8** Aerodynamic effects at a split flap as drawn in patent by Wright and Jacobs. (From Wright and Jacobs [10], figures 1–6.)

downstream. By this, the split flap can be regarded as virtually extending the airfoil – or generating a flow that would appear on an extended airfoil. The reduction of the trailing edge pressure on the lower side directly reduces the need of the flow to recover to the static pressure. The corresponding pressure gradients are therefore reduced and the boundary layer flow is capable to stay attached up to higher angles of attack.

### 5.1.3    Gurney Flap

The Gurney flap, named after the famous race car driver Dan Gurney, is similar to a miniaturized split flap, or Schrenk flap, with an orthogonal deflection to the surface.

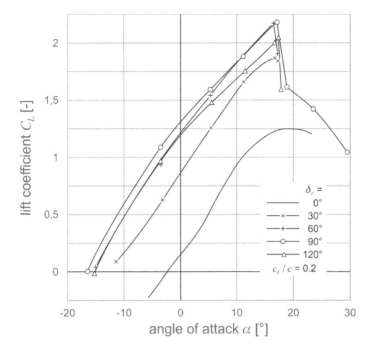

**FIGURE 5.9**   Lift coefficients of wing with [20% split] flaps vs. angle of attack. (According to Schrenk & Gruschwitz [11], figure 4(c).)

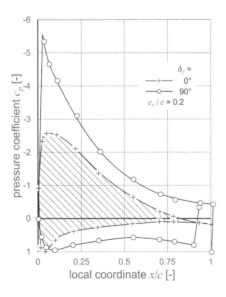

**FIGURE 5.10**   Pressure distribution about smooth wing (hatched) and about wing with flap. (According to Schrenk & Gruschwitz [11], figure 5.)

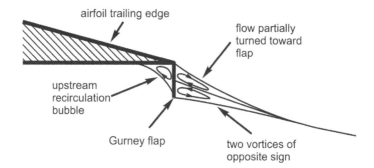

**FIGURE 5.11** Hypothesized trailing-edge flow conditions of the airfoil [...] with a Gurney flap. (Reproduction of Liebeck [12], figure 31.)

Dan Gurney (*1931) is one of the most famous American race car drivers. *He was the first of three drivers to win races in four different series (Sports Cars 1958, Formula-1 1962, NASCAR 1963, and Indy Car 1967).*[2]

*As owner of* [his own race car company] *All American Racers, he was the first to put a simple right-angle extension on the upper trailing edge of the rear wing. This device, called a Gurney flap, increases downforce and, if well designed, imposes only a relatively small increase in aerodynamic drag.*[2]

**PHOTO**   Joost Evers, Nationaal Archief[1]

The Gurney Flap on a Porsche 962 rear wing[3]

**Sources: Wikipedia**
[1]https://commons.wikimedia.org/wiki/File:Dan_Gurney_(1970).jpg
[2]https://en.wikipedia.org/wiki/Dan_Gurney
[3]https://commons.wikimedia.org/wiki/File:Andretti_962_HR7_gurney.jpg

Figure 5.11 reproduces the principal features of the flow field, as sketched by Liebeck [12]. A typical length of a Gurney flap is only a few percentages of airfoil chord. The Gurney flap is mostly seen as a permanent device, and mainly in automotive sports, but there are a few applications on aircraft, too.

The retractable Gurney flap goes back to a patent of Zaparka [13] who found that a much smaller device than a split flap, located at the trailing edge, already achieves a significant increase of the maximum lift coefficient. The assumed aerodynamic mechanism was basically the same artificial increase of the lower side pressure and the decoupling of upper and lower side trailing edge pressure as, at first glance, intended by the split flap. Although this patent is much older, even today, the movable device is more associated with the name of Dan Gurney despite the fact that he only used the rigid version.

The pure Gurney flap is not seen often on aircraft, especially not for lift increase. The additional drag in cruise prevents the use for the full application. Nevertheless, the principle has been adopted by McDonnel Douglas during the transition of the DC-10 to the MD-11 aircraft [14] in type of the so-called divergent trailing edge [15]. It differs from the pure Gurney only by filling up the upstream separation bubble. Moveable Gurneys, named Mini-TEDs, have been investigated and flight-tested on an A340 aircraft in the scope of the European AWIATOR project [16].

### 5.1.4   DROOP NOSE

Transferring the concept of cambering the airfoil from the trailing edge to the leading edge builds up a concept that orientates the leading edge at high angles of attack into the direction of the oncoming flow (Figure 5.12). Originally, this device was simply called "nose flap," but today the terminology "droop nose" is more common as the nose is drooped into the flow.

The basic concept goes back to Ludwig Bölkow. Originally, he proposed to droop a part of the leading edge to open a slot in the airfoil. Nevertheless, he also proposed the complete movement of the device [17]. Droop noses are commonly seen on military aircraft as the thin wings do not allow for more complex high-lift systems at the leading edge. On civil transport aircraft, they have been rediscovered more recently and are installed at the inboard wings of the Airbus A380 [18] and A350 [19].

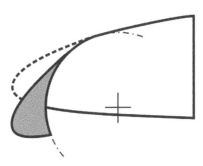

**FIGURE 5.12**   Principle of a droop nose.

Ludwig Bölkow (*1912–†2003) graduated in mechanical engineering at the Technische Universität Berlin. He joined the Messerschmidt AG during World War 2, where he was working on the aerodynamics of the Me 109 and the Me 263, the first jet-driven fighter airplane.

After the war, Bölkow founded his own aircraft company, Bölkow-Entwicklungen, later Bölkow GmbH, *which, with time, grew to the biggest aeronautics and spaceflight company, MBB (Messerschmitt-Bölkow-Blohm)*[2]. Today, the remains of the company are included in Airbus Helicopters and Airbus Defence and Space.

**PHOTO**   VDI-Archive[1]

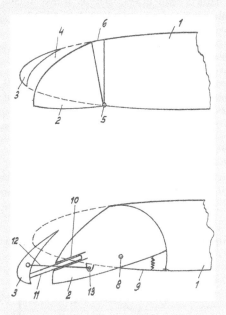

Drawings from Bölkow's original patent No. 694916 from 1939 [3].

**Sources:**
[1]https://www.vdi.de/mitgliedschaft/ehrungen/hall-of-engineers/
[2]https://en.wikipedia.org/wiki/Ludwig_B%C3%B6lkow
[3]Bölkow L (1939) Tragflügel mit Mitteln zur Änderung der Profileigenschaften, Patentschrift No. 694916, Deutsches Reichspatentamt.

Although the cambering of the airfoil by introducing a hinge is similar, the aerodynamic effects at first glance differ totally. While the plain cambering flap introduced a direct increase in lift coefficient at same angle of attack (see Section 5.1.1), for the droop nose, at similar conditions, the lift coefficient slightly decreases. A gain in maximum lift coefficient is achieved as the stall onset is shifted to higher angles of attack. Figure 5.13 shows a variation of the size and deflection angle of a droop nose at a NACA 0009 airfoil. In the linear range, a small degradation of lift coefficient is visible. The increment of the maximum lift coefficient is mainly dependent on the deflection of the droop nose, while the length mainly affects the suddenness of the stall onset.

The corresponding pressure distributions, shown in Figure 5.14, reveal that the drooping of the nose into the flow reduces the leading-edge suction peak and the associated pressure gradient. The reduction of the suction peak additionally reduces the risk to run into compressibility limits at significant high Mach numbers as experienced by transport aircraft even in the low-speed flight phase.

At the hinge of the droop nose, a second suction peak forms due to the locally increased curvature. This second suction area compensates for the reduction of the leading-edge suction peak with respect to the lift generation. Logically, the shift of suction gets increased with increasing deflection angle. The size of the droop nose has only a minor effect on the leading-edge pressure reduction. The size only affects the position of the compensating suction at the hinge. Nevertheless, the further downstream the hinge is, the more the stabilizing local pressure reduction makes the stall onset smoother.

Alternative studies on droop noses prefer to avoid the discontinuous curvature on the upper side and the building of a second suction peak. Such droop noses try to realize a continuous deformation of the airfoil shape in the leading-edge region. Establishing a curvature continuous upper side is expected to have a beneficial impact on the development of the boundary layer. Figure 5.15 shows the pressure distributions for increasing angle of attack around such a smart leading edge in comparison to the same airfoil without leading edge droop. The smooth leading edge reduces the gradient of the pressure rise past the suction peak and makes it more homogenous in comparison to the hinged version described before. An aerodynamically optimized shape of such a smartly bending droop nose can achieve a gain in maximum lift coefficient by about 25% [20].

The big challenge of smart droop nose devices is twofold. First, it is necessary to determine a structural skin concept that allows for the bending of the surface, but at the same time to carry the aerodynamic loads without too much deformation due to varying flight conditions. First attempts were made in the United Kingdom during the RAEVAM (Royal Aeronautical Establishment – Variable Area Mechanism) study. The first patent of a conforming mechanism has been assigned to Pierce [21]. Current investigations use tailored composite structures with variable stiffness and thickness, e.g., by Kintscher et al. [20]. Second, it is necessary to find a compliant mechanism to induce the morphing motion and support the aerodynamic loads at a minimum amount of weight. While this mechanism had a lot of moving parts, the tendency is to either reduce the number [22] or to use flexure hinges with superplastic materials [23, 24], which would reduce the maintenance significantly.

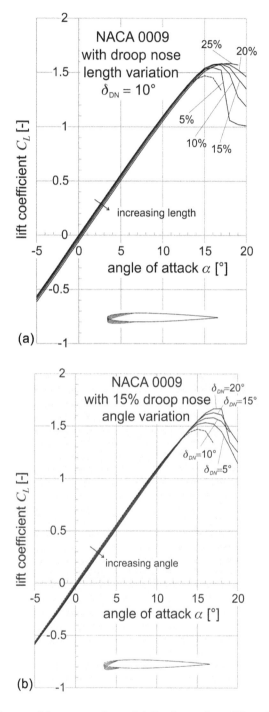

**FIGURE 5.13**  Impact of droop nose size and deflection angle on lift coefficient of a NACA 0009 airfoil: (a) increasing length; (b) increasing deflection.

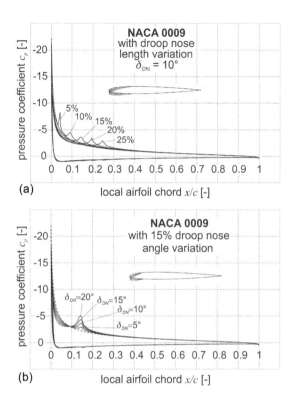

**FIGURE 5.14** Impact of droop nose size and deflection on the pressure distribution around a NACA 0009 airfoil: (up) size variation; (low) deflection variation.

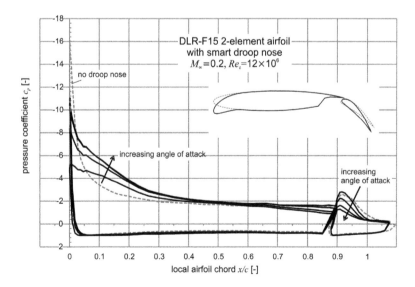

**FIGURE 5.15** Comparison of pressure distributions of a two-element high-lift airfoil with and without a smartly bending droop nose.

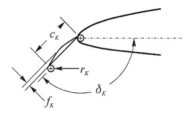

**FIGURE 5.16** Principal sketch of the nose split flap as drafted by Krueger. (Zoom of Krueger [25], figure 1.)

### 5.1.5 NOSE SPLIT FLAP (KRUEGER)

The nose split flap – today known as Krueger flap – is a device at the leading edge that is deflected from the lower side of the wing against the flow. Figure 5.16 reproduces the original sketch by Werner Krueger [25]. The hinge of the Krueger was

**THE MISSING KRUEGER FLAP PATENT**

Werner Krüger (*1910–†2003)[1] never claimed a patent on his "nose-split-flap." Instead, he published his results 1943 in form of the publicly available report 43/W/64 at the Aerodynamische Versuchsanstalt (AVA) in Göttingen.

Boeing partially used Krueger's flaps on its first jet aircraft, the 707 (entry into service 1958), the B727 and B737, and on full span on the B747 "Jumbo."[2]

In 1968, Frederick Thomas Watts filed the US patent no. 3363859A on a Krueger flap and further claimed license costs from Boeing. Boeing hired Werner Krüger and fought down the patent dispute, as Krüger was able to evidence his flap to be common knowledge.[3]

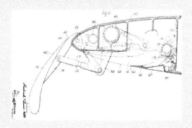

Drawing from Watts' original patent No. 3363859A from 1968[4].

**Sources:**
[1]https://en.wikipedia.org/wiki/Werner_Kr%C3%BCger.
[2]Rudolph PKC (1993) High-Lift Systems on Commercial Subsonic Airliners, NASA CR 4746.
[3]Meier HU (2011) Die Pfeilflügel Entwicklung in Deutschland bis 1945, Hamburg Aerospace Lecture Series, March 3, 2011.
[4]Watts FT (1968) Aircraft, US patent No. 3363859A, European Patent Office.

chosen to allow a nearly tangential attachment to the airfoil suction side. By this, the Krueger device extends the wing chord and introduces a reduced curvature at the original airfoil leading edge. As the leading edge in deflected position formed by the Krueger flap is hidden in retracted position, it can be subject to the design. Krueger reported a benefit in lift coefficient of $\Delta C_L \approx 0.5$ [25].

With the beginning of the age of jet-driven transport aircrafts, Boeing chose Krueger flaps for the leading edge as their principal high-lift device. The Boeing aircrafts B707, B727, and B737 were equipped with them on the inboard wing, and the B747 on the whole wing span – (Rudolph [26] p. 45ff). Figure 5.17 reproduces a sketch of a Boeing patent [27] that is likely associated with the B737 inboard Krueger flap. To improve the aerodynamic behavior, Boeing introduced the folding bull nose in addition. Here, the leading edge of the Krueger is folding away during retraction, which gives more degrees of freedom in the design and the sealing of the lower side in retracted position.

Werner Krueger tested the nose split flap in conjunction with a clean laminar wing airfoil and the same airfoil equipped with a trailing edge split flap [25]. Figure 5.18 shows the summary of this data. Focusing on the data without split flap (hollow symbols), it is seen that the slope of the lift curve is steeper with the deflected Krueger flap. In the linear range, the lift coefficient is significantly less than for the clean airfoil. An additional effect is seen for the curves with the deflected trailing edge split flap (solid and half-solid symbols) since the data there extends more towards low angles of attack. In this region, a strong non-linear behavior is seen in comparison to the measurements without Krueger flap (solid and half-solid circles). This nonlinearity is attributed to a full separation of the flow

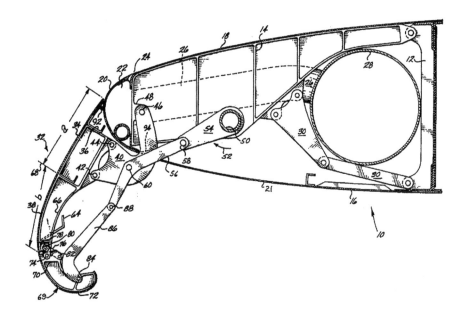

**FIGURE 5.17**   The folding bull-nose Krueger flap similar to the Boeing B737. (James & Hill [27], figure 1.)

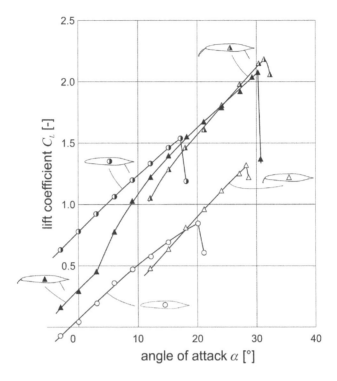

**FIGURE 5.18** Measurements of the lift coefficient of different nose split flaps with and without trailing edge split flap by Krüger. (According to Krüger [25], figure 3 left.)

on the lower side past the Krueger device. Due to the sharp leading-edge radius and the steep inclination of the Krueger panel, the flow does not reattach to the lower side of the wing, thus losing a significant contribution to the generation of lift. A second aspect not studied by Krueger is the flow in the forming cavity. The success of the Boeing aircraft may serve as evidence that this cavity is of minor importance for the design.

## 5.2  SLOTTED AIRFOILS

Before discussing the different types of high-lift systems with slots, it is necessary to look in more detail into the aerodynamic effects that make slotted airfoil so powerful in terms of lift generation. At the beginning of the application of slotted airfoils, the improved lift capability was fully attributed to the stabilization of the boundary layer. Two effects were mentioned that were thought to be responsible. First, the segmentation of the boundary layer into multiple sections makes every single boundary layer itself more resistive against the corresponding pressure rise. Second, the slots through the airfoil were attributed to re-energize the boundary layer by introducing a high-energy flow towards the weakened flow. While the first effect is in principle correct, it cannot explain the large potential of additional lift generation. The second effect, although still widely distributed, is simply not the mechanism sought as we will see later in this section.

The first to write down the aerodynamic effects that make slotted high-lift devices the most powerful means of passive lift generation was A.M.O. Smith [28]. He described the effects of the high-lift system as five major effects. Woodward and Lean stated in 1993 "... the aeronautical community should, for 54 years, have a totally incorrect view of the physical principles underlying its operation. Until the publication of A.M.O. Smith's classic paper in 1972[3], it was accepted widely that the slot performs as a boundary layer control device – and this is not true" – (Woodward and Lean [33] p. 1-2.).

## 5.2.1 The Theory of Multi-Element Airfoil Aerodynamics

It was one major achievement of A.M.O. Smith to describe the effects that make slotted airfoils such powerful concepts of high-lift systems. He described the first three effects of the change of the pressure distribution by potential flow theory to emphasize that the additional lift is not majorly attributed to the stabilization of the boundary layer but the positive influence of multiple lifting bodies on the pressure distribution of each other.

### Effect #1: Slat Effect (Figure 5.19)
In the vicinity of the leading edge of a downstream element, the velocities due to circulation on a forward element, for example, a slat, run counter to the velocities on the downstream element and so reduce pressure peaks on the downstream element.

**(Smith [28] p. 518)**

This effect is mainly responsible for the increase of the angle of attack at which an airfoil stalls. Due to the reduction of the suction pressure, the following pressure rise and the pressure gradient are largely reduced and the airfoil can be brought to a higher angle of attack to have the same pressure rise as it would have in an isolated manner.

Due to the reduction of the suction peak and the upper side low pressure, the main airfoil produces less lift at the same angle of attack. Liebeck [29] showed by simplified potential flow computations that the loss in lift on the main airfoil is nearly equal to the lift corresponding to the circulation of the upstream vortex. The total circulation needed to bend the flow to achieve the Kutta-condition at the trailing edge of the airfoil is unchanged. This behavior becomes even clearer if it is remembered that the

**FIGURE 5.19**  Schematics of the slat effect: a lift generating vortex placed ahead of an airfoil.

**FIGURE 5.20**   Schematics of the circulation effect: a lifting vortex placed downstream of an airfoil.

suction pressure is the generating force to compensate the centrifugal forces of the flow following the airfoil curvature. Thinking of an upstream element as a guiding vane element, the upstream circulation takes over some of the flow turning and the suction force needed to keep the flow along the airfoil contour is less.

Another interpretation that the effect results from a reduced angle of attack due to the downwash of the upstream element is less exact. A change of the angle of attack would act globally on the airfoil, while the effect seen is reducing with the distance from the upstream element. And, it is relative to the ratio of the circulations of the two elements. For example, for a slat device, the effect can only be seen locally at the leading edge of the main wing, while the effect of a wing on a flap is more global.

EFFECT #2: CIRCULATION EFFECT (FIGURE 5.20)
The downstream element causes the trailing edge of the adjacent upstream element to be in a region of high velocity that is inclined to the mean line at the rear of the forward element. Such flow inclination induces considerably greater circulation on the forward element.

**(Smith [28] p. 518)**

Altering the condition at the trailing edge directly influences the circulation of the airfoil as already seen in Chapter 2. The additional circulation of the airfoil needs to compensate exactly the velocity component induced vertically to the trailing edge. Therefore, the overall lift is not only increased by the circulation of the lifting vortex but also by the rising circulation of the airfoil. The slot, therefore, acts as an ampli-fier for the lift generated e.g., by a flap. This is the main reason why a slotted flap produces significantly more lift than a plain flap at the same angle of attack and flap deflection, even when the plain flap flow is attached.

It has to be highlighted that the maximum lift is not increased by this effect but only the lift at the same angle of attack. Since the maximum lift is coupled to a limit-ing pressure distribution – either by viscous or by compressibility effects – the circu-lation effect is responsible that the limiting pressure distribution would be obtained at lower angles of attack. Therefore, purely by the circulation effect, the complete airfoil would stall at a lower angle of attack but at a similar (partial) lift coefficient as without the downstream element.

It is important to understand that slat and circulation effects appear at every slot of the multi-element airfoil, and it is only a question of upstream and downstream. A main wing airfoil introduces a slat effect on the flap and a circulation effect on a slat element in a similar way. And, it is therefore important to understand that these

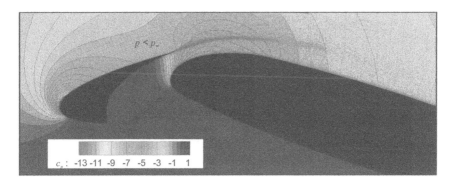

$p < p_\infty$

$c_p:$  -13 -11 -9  -7  -5  -3  -1   1

**FIGURE 5.21**   Dumping effect visualized by a slat flow field – the trailing edge of the slat is in the accelerated flow region of the main airfoil.

two effects do not appear isolated but are always coming together. And one effect alters the other effect, e.g., a strong slat effect by the upstream element reduces the circulation on the downstream element. The so reduced circulation effect by the downstream element thus reduces the circulation of the upstream element. By this, the slat effect is reduced leading to an increased circulation of the downstream element, and so forth. The real situation at a multi-element airfoil is therefore a balance between these two effects.

### Effect #3: Dumping Effect (Figure 5.21)

Because the trailing edge of a forward element is in a region of velocity appreciably higher than freestream, the boundary layer "dumps" at a high velocity. The higher discharge velocity relieves the pressure rise impressed on the boundary layer, thus alleviating separation problems or permitting increased lift.

**(Smith [28] p. 518)**

Although this effect directly affects the pressure distribution, the action of delaying the stall of the upstream element is achieved by stabilizing the boundary layer. Referring back to Section 2.3, the importance of the trailing edge pressure was mentioned when looking into the dimensionless canonical pressure distributions. It is perfectly clear that, by lowering the trailing edge pressure, the minimum pressure can be reduced much more and the pressure gradient can be much higher.

Of course, also circulation effect and dumping effect always come together. Only in combination, the increase in maximum lift by adding a slotted flap to an airfoil is observed at only a slightly reduced angle of attack for stall.

It is worth to note that the basic findings would have been available much earlier. Already Alfred Betz [30] drew a diagram – reproduced in Figure 5.22 – on the change of the pressure distribution for two airfoils in sequence based on the work of Lachmann, in which the three effects are shown separately. The solid line (1) shows the pressure distribution of the airfoils if they were not interfering. The dashed line (2) at the downstream airfoil shows the slat effect with the reduction of the suction peak. The dashed line (2) at the upstream airfoil shows the reduction of the trailing

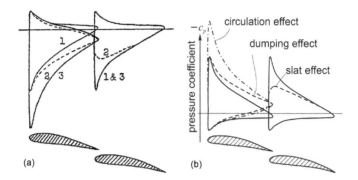

**FIGURE 5.22** Change in pressure distribution due to mutual influence of both wing sections. 1. Undisturbed pressure distribution (fine line). 2. Disturbed pressure distribution with unchanged angle of attack (dash line). ((a): Original from Betz [30], figure 5; (b): Reproduced with inverted axis)

edge pressure due to the dumping effect. The effect that Betz missed was the circulation effect. His pressure distribution sketched as solid line (3) was meant to be the pressure distribution at a higher angle of attack where the pressure gradient is the same as for the initial airfoil. Instead, the solid line (3) at the upstream airfoil together with the flap pressure distribution for (2) would show the increase of circulation due to the circulation effect. Finally, the only property not included by Betz is the movement of the stagnation points due to the changed circulation.

Nevertheless, Betz – a colleague of Ludwig Prandtl – concluded: "Here the whole of the front section lies in a field of increased velocity and is thus able to produce a greater lift, since the lift is proportional to the square of the velocity" – (Betz [30] p. 9f) – and "The slots convey new energy to the marginal layer of air retarded by friction on top of the wing, thereby increasing its velocity and thus preventing the accumulation of dead air. The air stream flowing out of the slot acts like the jet from a syringe and reinforces the air stream on top of the wing in carrying away the dead air" – (Betz [30] p. 11). Besides the already mentioned misinterpretation of slots as a boundary layer control device, it introduced a long-lasting misunderstanding of the flow through the slot of a high-lift system to be similar to a nozzle flow or a jet. Still 1985, in the standard textbook "Fluid Dynamic Lift" by Hoerner & Borst [31], it is stated that the "Boundary layer control by means of a slot (such as in slotted trailing edge flaps) is based on the concept of injecting momentum into a 'tired' boundary layer" – (Hoerner & Borst [31] p. 6-1). In fact, the flow characteristic is completely opposite of this and the explanation as a nozzle leads to the wrong conclusions in terms of designing a slotted airfoil.

First of all, the flow coming out of the slot is not a high-energy jet. The fluid outside the boundary layers and shear layers has the same stagnation pressure as far as isentropic flow is assumed. As there is no energy source, it will never have a higher total energy than the oncoming flow. In fact, due to the wall boundary layers inside the slot, the energy will be slightly less than the oncoming flow.

Second, this is not a nozzle flow. A major flow characteristic that is associated with a nozzle flow is the increase of flow velocity with reduced cross-section. This

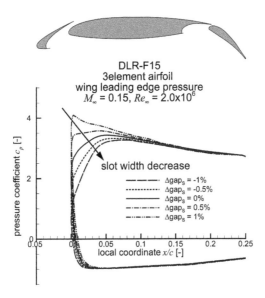

**FIGURE 5.23**   Impact of gap reduction on the suction pressure peak of the main airfoil element of the DLR-F15-VLCS airfoil.

is needed to fulfill the continuum equation in a ducted flow. For slotted high-lift systems, this would impose that the flow velocity through the slot should increase with reduced gap size. Figure 5.23 shows the development of the pressure coefficient in the gap between a slat and a wing for different gap sizes. It is clearly seen that the suction pressure at the wing drops with reducing gap indicating a reduction of the velocity of the fluid through the gap. The reason for this behavior is mainly the slat effect itself. With reducing distance of the upstream circulation, the slat effect gets stronger and leads to the seen drop of the wing leading edge suction. The main mistake made when thinking of a slot like a nozzle is that there is no constraint on the mass flow through the slot. By reducing the gap size, the mass flow passing through the slot is reduced twice: by the smaller cross-section and the decreased velocity in the smallest cross-section at the end of the slot.

### EFFECT #4: OFF-THE-SURFACE PRESSURE RECOVERY
The boundary layer from forward elements is dumped at velocities appreciably higher than freestream. The final deceleration to freestream velocity is done in an efficient manner. The deceleration of the wake occurs out of contact with a wall. Such a method is more effective than the best possible deceleration in contact with a wall.

**(Smith [28] p. 518)**

To explain this effect, it is necessary to look into the deceleration of a free wake flow. In adiabatic flow, the relation of pressure and velocity along a streamline is covered by Bernoulli's equation

$$p + \frac{1}{2}\rho U^2 = \text{const.} \tag{5.4}$$

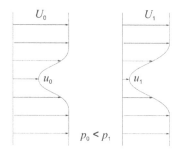

**FIGURE 5.24**   Velocity distributions of a wake in adverse pressure gradient.

Figure 5.24 shows the principle velocity distribution of a free wake flow in adverse pressure gradient. For some stream line outside the wake, the velocities shall be denoted by capital "U," for the core of the wake with lower case "u."

At both positions, the Bernoulli equation is valid if no outer energy is introduced

$$p_0 + \frac{1}{2}\rho_0 U_0^2 = p_1 + \frac{1}{2}\rho_1 U_1^2$$

$$p_0 + \frac{1}{2}\rho_0 u_0^2 = p_1 + \frac{1}{2}\rho_1 u_1^2.$$

(5.5)

Assuming incompressible flow

$$\rho_0 = \rho_1 = \rho_\infty$$

(5.6)

and using the definition of the canonical pressure coefficient from eq. (2.72) with the reference velocity $U_0$, the ratio of the velocities at the core of the wake and the outer flow is obtained as

$$\frac{u_1^2}{U_1^2} = \frac{\left(u_o^2 / U_o^2\right) - \bar{c}_p}{1 - \bar{c}_p}.$$

(5.7)

It shall be remembered that the value $\bar{c}_p = 1$ corresponds to the stagnation of the outer flow ($U = 0$). It is seen from eq. (5.7) that the core velocity of the wake becomes zero even before the stagnation of the outer flow. The flow in the core of the wake can achieve stagnation even before the outer flow. This is even more pronounced if the velocity deficit in the wake is high and the pressure rise happens on a short distance, so that viscous entrainment effects are not able to accelerate the core flow.

This situation can happen on a high-lift system in the wake of an element passing the suction side of a following element, especially over highly deployed slotted flaps as seen in Figure 5.25. The very pronounced wake of the main airfoil resulting from the relieved boundary layer disposed over the flap in the accelerated region is subject to the pressure rise over the flap. It is seen that the wake widens and the core gets to stagnation at the third depicted position. After this position, the flow is subject to

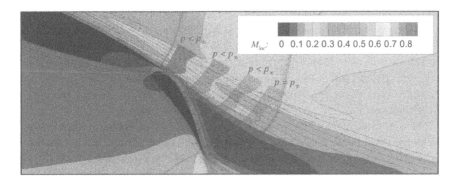

$M_{loc}$:   0  0.1 0.2 0.3 0.4 0.5 0.6 0.7 0.8

$p < p_\infty$

$p < p_\infty$

$p < p_\infty$

$p = p_\infty$

**FIGURE 5.25**   Flow pattern above a highly deflected flap at high angles of attack.

further pressure rise that is then only achieved by pressure diffusion as the flow itself has no further kinetic energy. In conclusion, there exists a dead-water above the flap that additionally can be seen to act as a displacement body.

Anyhow the lift generation by the multi-element airfoil is defined by the circulation around the wing and the flap. For both, the flow is still attached. Therefore, the determining condition for the circulation – the Kutta-condition – is not affected. Both elements produce lift in the same manner as without the stagnation or recirculation area above the flap. Therefore, releasing a fluid regime that has experienced viscous losses as a free shear flow into an adverse pressure gradient is more effective in terms of lift generation.

Nevertheless, it has to be mentioned that the existence of the stagnation has, of course, an impact on the airfoil drag. A flow field as shown in Figure 5.25 experiences a high drag coefficient. The pressure diffusion into the stagnating flow imposes an increased level of pressure drag. Therefore, such a situation is acceptable for a landing configuration where drag is of minor importance. For a take-off configuration, such a situation should be avoided.

### EFFECT #5: FRESH BOUNDARY LAYERS
Each new element starts out with a fresh boundary layer at its leading edge. Thin boundary layers can withstand stronger adverse gradients than thick ones.

**(Smith [28] p. 518)**

The separation criterion by Stratford's formula (eq. (2.78)) indicates for a constant risk of separation the relation

$$x \frac{\partial \overline{c}_p}{\partial x} = const. \tag{5.8}$$

with the other parameters, e.g., the Reynolds number, being constant. This implies that the flow around an airfoil element having half the chord is able to withstand the double pressure gradient. This is also one reason why the flow around short slats is relatively stable to separation. On the other hand, it becomes clear that the fresh boundary layer is not the major effect for the high lift coefficient potential related to separation of the main wing element, as here the chord length is reduced by only about 15% compared to a clean airfoil. This would only allow for a lift coefficient increase by not more than 13%. On the other hand, a multi-element airfoil is able to achieve a lift coefficient that is about twice as high as a clean airfoil. So, the fresh boundary layer effect is a minor add-on but not the reason for the high potential of lift generation of slotted airfoils.

There is a sixth effect mentioned by Smith that is not attributed to the increase of lift by the slots, but to a new risk of premature stall onset. This effect is concerning the possible mixing of the shear layers. Smith states: "It is our observation [...] that gaps between airfoil elements should be so large that wakes and boundary layers do not merge for, if they do, early separation will set in" – (Smith [28] p. 516).

## 5.2.2 SLATS

A slat is a device positioned at or deployed from the leading edge of the wing (Figure 5.26). The first slat by Handley-Page was a rigid device forming a slot at the front of the wing. As discussed before, it mainly makes use of the slat effect and, therefore, increases the stall angle by preventing the separation at the main wing. Typical slat devices achieve a relative chord length of about 12–20% of the local wing chord. By this, they have a low separation risk in terms of turbulent boundary layer separation on their own due to the short length and the severe dumping effect due to the highly accelerated flow at the leading edge of the main wing. The major risk for separation at a slat is dedicated to either laminar separation bubbles at lower Reynolds numbers (below $Re \approx 3 \times 10^6$) or compressible effects. The strong acceleration around the slat leading edge can achieve supersonic speeds. In the following, interaction between a forming sonic shock and the boundary layer can lead to a local separation. In the ultimate case, the Mayer limit discussed in Section 2.2 poses an upper limit to the suction. To underline this, a look is taken at the development of the minimum pressure coefficient at the leading edge of a slat.

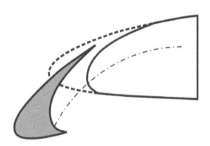

**FIGURE 5.26**  Principle of a slat.

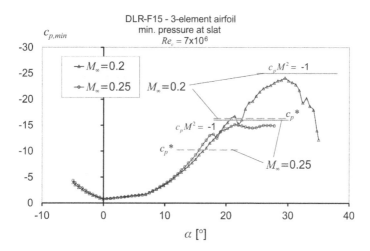

**FIGURE 5.27** Evolution of minimum pressure coefficient at a slat over angle of attack for two onflow Mach numbers obtained by experiments with the DLR-F15 3-element airfoil. (Reduction of Figure 2.11 and from Wild [32], figure 13.)

Figure 5.27 shows the variation of the minimum pressure coefficient of the slat with angle of attack. The figure is a repetition of Figure 2.11 of Section 2.2 with emphasis on the two higher onflow Mach numbers where the flow around the slat enters the transonic regime. It is seen, that passing the sonic boundary introduces a dip in the curve. In this case, the shock is not yet introducing a stall limit, but the introduced losses reduce the circulation around the slat. In the following with increasing angle of attack, the pressure level drops close to the Mayer limit. It is worth to note that at the higher free stream Mach number, even reaching the Mayer limit does not necessarily limit the lift generation of the whole airfoil system. Only the suction limit of the slat is limited, while the rest of the airfoil continues to produce lift until reaching its own stall onset.

Figure 5.28 shows the streamlines of the flow around a typical slat of today's transport aircraft at a high angle of attack close to stall. The high circulation around the complete multi-element airfoil pushes the stagnation points of the main wing and the slat far downstream. For the slat, the stagnation point can move downstream as far as the lower trailing edge. The flow stream lines are highly bent so that the local flow direction can get perpendicular to the free stream flow direction. Also, for the main wing, the local flow direction is highly inclined compared to the onflow direction. In the rear of the slat, the so-called slat cove, a recirculation area is forming that becomes smaller with increasing angle of attack. The figure also indicates the area of flow with viscous losses, namely the wake of the slat and the wing boundary layer. The slat boundary layer remains very thin and is almost not visible. The slat wake is approximately double the size of the wing boundary layer and along the wing, a small area of flow remains that is not affected by viscous effects. Smith [28] states "that gaps between airfoil elements should be so large that wakes and boundary layers do not merge" at least not before the end of the main wing, "if they do, early separation will set in" – (Smith [28] p. 516).

The most prominent aircraft equipped with a rigid slat was the Fiesler Fi 156 "Storck," which is renowned as an aircraft with extreme high-lift and low-speed performance. The stall speed of the aircraft was as low as 45 km/h (23.15 kts) at a lift coefficient of $C_{L,max} = 3.9$. Equipped with a single 240 hp piston engine, the take-off field length was as low as 155 m and the best climb rate at about 5 m/s.

Rigid slat mounted at the leading edge of the wing of the Fi156 "Storck"

**Reference**
Hoerner S (1938) Erfahrungen und Messungen am Langsamflugzeug Fi 156 "Storch," In: Bericht über die Tagung Auftriebs- und Lesitungssteigerung in Göttingen am 14. Juni 1938, Bericht 099/006.

Woodward and Lean reported on a large study to detect the optimum position of a slat device within the British National High-Lift Program [33]. Based on numerous measurements of a slat at different positions and different deflection angles, the impact of the slat setting on the maximum lift coefficient was studied in detail. The graphs reproduced in Figure 5.29 show iso-lines of the maximum lift coefficient based on the position of the slat trailing edge in relation to the wing leading edge. The coordinate is chosen with the origin placed on the clean wing chord at the lateral

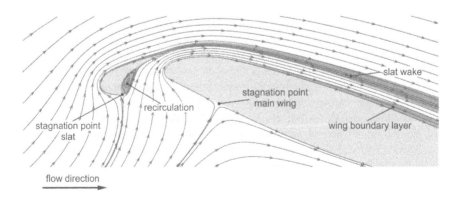

**FIGURE 5.28** Flow field around a slat at high angle of attack.

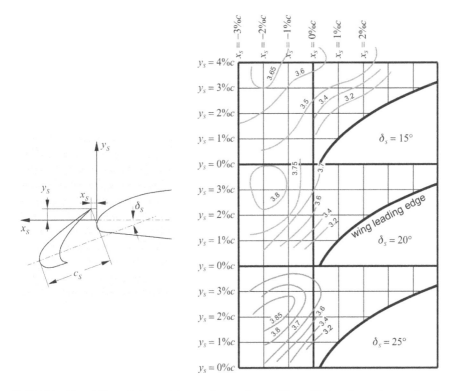

**FIGURE 5.29**  Effect of slat position on the maximum lift coefficient depending on the slat deflection angle. (According to Woodward & Lean [33], figure 20.)

position of the main element's leading edge. The blanked area shows the main airfoil contour so that positioning of the slat in this area is not possible. The grids of the graphs are in percentages of clean wing chords.

The more global trend to see is that the slat position exhibits a clear maximum for maximum lift coefficient. This maximum is closer to the main airfoil the larger the slat deflection angle is. Additionally, the optimum position is with a negative overlap, meaning that the slat trailing edge is positioned upstream of the main airfoil leading edge.

Second, the maximum lift coefficient increases with increasing slat deflection. Third, in the area between the optimum position and the wing contour, the iso-lines are almost parallel to the wing contour. The airfoil combination loses 10–20% of the maximum lift within a gap variation of about 2% of the clean airfoil chord. This indicates that the gap width is the major parameter for the slat positioning and the position along the contour is of minor importance.

The reason for this strong sensitivity can completely be explained with the effects described in Section 5.2.1. It has to be remembered, that for the slat device, the higher maximum lift coefficients are achieved at even higher angles of attack. Coming from far, bringing the slat closer to the wing increases the maximum lift coefficient by preventing the stall through the slat effect up to the optimum position. Getting closer, the risk of merging of the wing boundary layer and the slat wake increases, especially due to the long distance up to the wing trailing edge. The strong reduction of the maximum lift coefficient at gaps closer than the optimum can be attributed to this. Although the sensitivity in the overlap direction is less pronounced, it is also obviously related to the slot effects. With the slat too far upstream, the slat effect is not acting as much. With the slat too far downstream, the trailing edge is no longer in the optimum position for the dumping effects, i.e., in an area with already decelerated flow past the suction peak of the main wing.

There is another aspect of slats placed too far downstream that has to be taken into account. Depending on the deflection angle, at positive overlaps the slot between the slat lower side and the wing upper side can get convergent-divergent, i.e., the smallest gap value is no longer at the trailing edge of the slat. In this case, the slot past the smallest cross-section acts like a diffuser. If this diffuser is too steep, the deceleration of the flow and the therewith combined weakening of the main element boundary layer can also be responsible for an early onset of the main element separation.

### 5.2.3  SLOTTED KRUEGER FLAPS

The slotted Krueger flap is a variant of the device discussed before. Boeing introduced gapped Krueger devices in conjunction with a variable camber (VC) on the outboard wing of the Boeing B747. Figure 5.30 reproduces a diagram of a most probably associated Boeing patent [34]. Rudolph states that "the 747 VC Krueger has only a small performance advantage in maximum lift over a three-position slat, and its complexity [due to numerous and heavy support structure] cannot be justified" – (Rudolph [26] p. 84).

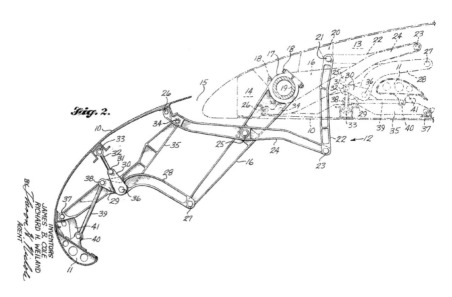

**FIGURE 5.30**  The folding bull-nose variable camber Krueger flap with slot of the Boeing B747 outboard wing. (From Cole & Weiland [34], figure 2.)

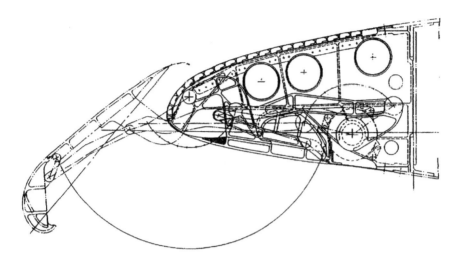

**FIGURE 5.31**    Slotted folding bull nose Krueger as used in the Boeing B757 Hybrid Laminar Flow Control (HLFC) flight test. (From Rudolph [26], figure 3.3.)

This concept receives a revival in the scope of the research on laminar wing technology [35]. Rudolph reports on the reasons to choose such a device for a flight test with hybrid laminar flow control on a Boeing B757, sketched in Figure 5.31: "A slat was ruled out because its aft step would have caused boundary layer transition and would have made laminar flow downstream of the front spar impossible. The insect protection requirement called for the Krueger to extend above and below the fixed leading edge, or a position that resembles that of a deployed slat. Therefore, the Krueger had to have a slot and a very large bull nose" – (Rudolph [26] p. 90). But it was also mentioned that "there has been no effort to develop the fixed-camber Krueger into a device that has characteristics similar to that of a slat, except for the work done on the 757 hybrid laminar flow experiment" – (Rudolph [26] p. 152). One of Rudolph's conclusions on the further development of leading-edge high-lift systems was "... to explore and establish the potential of fixed-camber Krueger flaps–how close they can come in performance to the best slat or variable camber Krueger ..." – (Rudolph [26] p. 152). Especially in the context of laminar wing technology, the slotted Krueger flap is the most promising concept to achieve the high maximum lift potential of a slotted leading-edge device together with the contamination shielding properties.

Recent European research in the context of high-lift systems for laminar wings further matured this concept [36–38]. It has been shown that the high-lift performance obtainable by a laminar wing with Krueger device does not fall too short in comparison to a turbulent wing with a slat. Figure 5.32 compares the measured lift characteristic of a natural laminar wing with a slotted Krueger device to the turbulent reference wing with a slat, both implemented in the DLR-F11 half model. Although both wings have a different sweep angle, they are sized to be comparable to the aircraft requirements. This is of major importance as the high-lift performance may

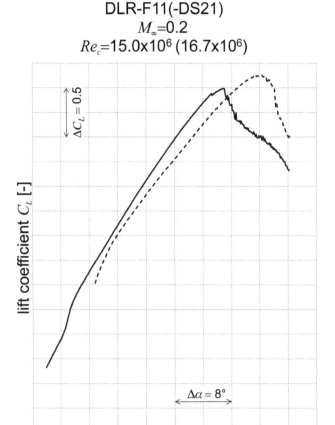

**FIGURE 5.32**  Comparison of lift coefficient vs. angle of attack for a Krueger flap and a slat on comparable wings measured on DLR-F11(-DS21). (from Wild [39], figure 5.)

size the needed wing area. An increased wing area would spoil the drag reduction benefits of the laminar wing technology.

### 5.2.4  SLOTTED FLAPS

Handley Page was the first to implement slots into wings for increasing lift. As for the slat device, he was also the first to promote slot openings at the flap to prevent the upper side separation at high deflection angles [40]. The simple slotted flap based on a slightly dropped hinge point of the flap dates back as early as the patents for slats. Figure 5.33 illustrates the basic principle together with a sketch of the first patent issued by Handley Page [40].

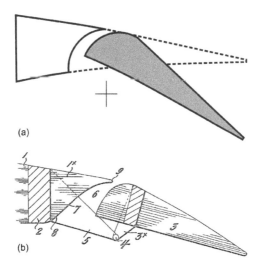

(a)

(b)

**FIGURE 5.33** Simple slotted flap forming a gap when deflected positively: (a) principle; (b) sketch from patent of Handley Page. (From Handley Page [40], figure 3.)

Due to the significant benefit for lift generation and the simple mechanics, this device type became very popular. Even today, it is often seen on lighter aircraft of the general aviation. The slot opening is able to delay the stall on the upper side of the flap up to 45° deflection angle, compared to 30° without slot at the plain flap (see 5.1.1).

Very detailed investigations for such flaps were performed based on the NACA 23012 airfoil [41, 42]. Figure 5.34 shows the arrangement of one specific slotted flap measured. The hinge point is located slightly below the airfoil and aft the flap leading edge. The ideal path of the flap leading edge is nevertheless not an arc. The nose

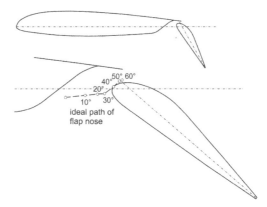

**FIGURE 5.34** Optimal flap position of a NACA 23012 airfoil with 25.66% slotted flap. (According to Wenzinger & Delano [41], figure 1(a).)

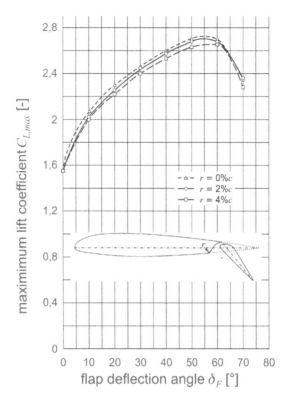

**FIGURE 5.35**  Effect of slot entry radius on $C_{L,max}$. (According to Wenzinger & Harris [42], figure 17.)

point is the most forward point on the deflected flap contour and, therefore, changes its location along the circumference of the leading-edge radius.

Figure 5.35 exemplarily shows some measurements of the maximum lift coefficient depending on the flap deflection angle, here with varying radii of the lower wing edge. The flap shows progressively increasing maximum lift coefficient up to deflection angles of about $\delta_F = 60°$. Corresponding pressure data [41] shows that the flap flow is attached up to a deflection angle up to $\delta_F = 40°$. At higher deflection angles, the flap's upper side is separated to a wide extent. An interesting – and for slotted flaps typical – behavior is seen for $\delta_F = 50°$ flap deflection.

Figure 5.36 reproduces the variation of the pressure distribution on wing and flap for the NACA 23012 airfoil with a slotted flap of $c_F/c = 25.66\%$ set at high deflection angle. While for the lower angle of attack, the flap flow is largely separated at $50°$ deflection, at the higher angle of attack of $\alpha = 12°$ the flap flow is attached. This behavior is directly related to the slat effect of the wing towards the flap. As the wing produces more lift, the flap is de-loaded as the suction level is reduced. This implies that a slotted flap produces less lift with increasing angle of attack. The flap "feels" a reduced onflow angle at an increased angle of attack of the whole airfoil due to the downwash of the preceding wing. Therefore, a flap that is separated at low angles of

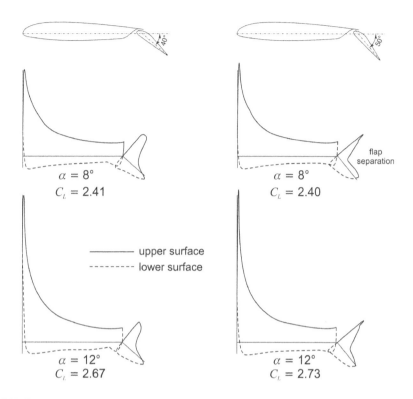

**FIGURE 5.36**   Pressure distribution on the N. A. C. A. 23012 airfoil with a 25.66% slotted flap, at high angles of attack. Flap set at 40° & 50° (According to Wenzinger & Delano [41], figures 7 & 8, resp.)

attack may reattach at high ones. Practice has even shown that the highest maximum lift coefficients can be achieved with flaps reattaching just at the maximum lift condition. In this case, they contribute the most to the airfoil lift coefficient.

### 5.2.5   FOWLER FLAPS

Besides the slat, the Fowler flap named after the inventor HD Fowler was the big invention that made flying possible as we know it today. It was the significant lift improvement that achieves the necessary gain in lift coefficient for the low-speed regime to enable efficient transonic wing design as the basis of modern air transport.

What is sometimes overlooked is that there are two types of Fowler flaps. The first type – HD Fowlers' first invention – only targeted an increase of the wing area [43]. What is known today as the Fowler flap is the second type, which increases wing area, wing camber, and opens a slot [44]. Figure 5.37 shows the principal arrangement of a Fowler flap. In contrast to the slotted flap discussed before, the Fowler flap shows a much larger displacement in the downstream direction, a much longer shroud area, and – if applied by a simple rotation – the rotation center located much more off the lower wing surface.

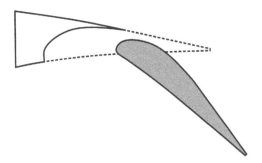

**FIGURE 5.37**   Principle of a Fowler flap.

After presenting this device to the public, numerous explanations for the big increase in lift coefficient were around. Eight years after publishing his patent, Fowler saw a need to explain his device [45]: "The combined use of variable area, camber, angle of attack and of the boundary layer, requires only an auxiliary aerofoil that is extended or retracted as the occasion demands. ... The major individual increase in maximum lift results from the gap, which appears to increase nearly directly with the chord of the fowler. Variable area is the next most effective factor, with camber a close third. However, by virtue of the principle involved, the combination of area and camber actually contributes a greater increase than the gap. It is apparent that all three factors have a powerful influence in contributing to the maximum lift, and increase with enlargement of the flap chord" – (Fowler [45] p. 247).

In the same publication, Fowler introduced also one specific standard that is still followed in high-lift aerodynamics: "As a high lift device is generally used to increase the high speed of a machine keeping a given stall speed, it is proper to consider the lift in terms of the normal wing area, with fowler retracted" – (Fowler [45] p. 248). This implies that the wing area increase is accounted for in the lift coefficient. In Figure 5.38, this influence is shown based on measurements made on a Clark Y airfoil with three flaps of different sizes as reported by Platt [46]. The measured maximum lift conditions are referenced either to the retracted wing area or to the sum of the wing and the flap area. Still, the lift increase seen by the slot alone is still very significant ($\Delta C_{L,\max} \approx 0.9$), as this is the effect when adjusting the wing area to the deflected situation. But this relation also indicates that the effect of the gap reduces with increasing flap size and a good size for a flap making as much as possible use of the gap effects would be at $c_F/c = 25\% \div 30\%$.

An implication of the large movement in chordwise direction is the need for more complicated mechanisms to place the flap at the desired position. The challenge is to find a proper deflection path enabling good aerodynamic properties not only for the highest deflection but also at intermediate positions. As seen in Chapter 4, the requirements for a good take-off and a good landing setting of the high-lift system differ substantially. This is accounted for by placing the flaps for these different targets at different positions.

Since the degrees of freedom for the layout of the kinematics allows for specific flap settings, it is worth to assess the sensitivity of the flow towards the flap

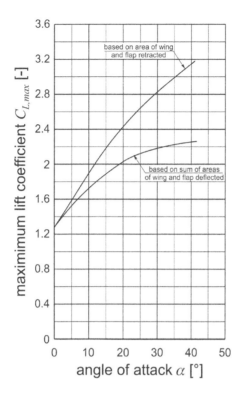

**FIGURE 5.38**  Maximum lift coefficient potential of a Fowler flap depending on reference area definition. (From Platt [46], figure 5.)

positioning. Woodward and Lean [33] report detailed investigations of the dependency of the maximum lift coefficient. Similar to the discussions made for the slat, Figure 5.39 shows iso-lines of the maximum lift coefficient depending on the flap position for two different flap deflections for the L1T2 3-element airfoil.

The introduced coordinate system defines the flap position by the front vertical and upper horizontal tangent to the flap contour. The origin is at the trailing edge of the main wing, which would be the spoiler trailing edge of a real wing. They report iso-lines of equal maximum lift coefficient for two different flap deflection angles. For the lower deflection angle of $\delta_F = 20°$ (Figure 5.39 upper right), representative for a take-off setting, the strongest sensitivity is seen in a diagonal direction towards the wing trailing edge, which is equivalent to the gap size of the slot between wing and flap surface. Positioning the flap closer to the wing trailing edge increases the slat effect of the wing onto the flap, and thereby reduces the flap loading. This, in turn, reduces the circulation effect upstream from the flap onto the wing. Putting the flap more downstream also reduces the circulation effect, and thereby the lift potential. As the flap is deflected only moderately, the flap itself is not yet prone to separation on its own and the slat effect is still sufficient to prevent a flap separation.

This behavior changes at the higher deflection angle of $\delta_F = 40°$ (Figure 5.39 lower right) that is representative of a landing setting. The optimum flap gap is at a

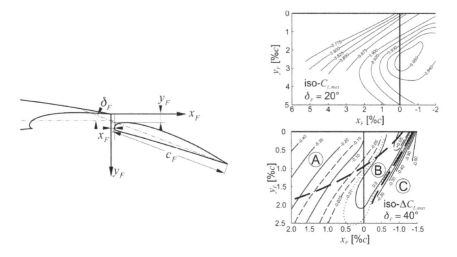

**FIGURE 5.39**  Sensitivity of maximum lift coefficient and the flap flow on the flap position: (left) coordinate system; (right upper) low flap deflection $\delta_F = 20°$; (right lower) high flap deflection $\delta_F = 40°$ (According to Woodward & Lean [33], figures 7, 24, and 25a.)

significantly lower value than for the take-off setting. The graph indicates the boundary of attached flap flow by thick dashed lines. The region marked with (A) denotes fully attached flow over the flap. In area (C), the flap flow is fully separated. In area (B), the flap flow is detached over a significant range of incidences. Woodward and Lean report "for some flap positions [...] the flow over the flap is separated at low incidence when the wing wake is thin, but, at higher incidence as the wing wake thickens, the flap flow suddenly re-attaches producing a characteristic non-linear increase in lift and negative pitching moment, and a reduction in drag. The incidence at which the re-attachment occurs gets higher as the flap moves away from the wing, until ultimately the flow fails to reattach at all, and a large decrement in maximum lift results" – (Woodward and Lean [33] p. 1-10). This reflects the higher request for the slat effect from the wing onto the flap to prevent the flap separation at its high deflection angle. By shifting the flap away from the wing, the slat effect weakens and the flap separation occurs very suddenly. The close location of the optimum position to the separation boundary underlines that the best maximum lift coefficient is achieved for a flap close to separation. As discussed for the simple slotted flap, this implies that the flap will show a separated flow at angles of attack lower than that for the maximum lift coefficient. The separation for closer gap values is likely to be attributed to the mixing of the wing wake and the flap boundary layer, similar to what has been discussed for the slat. Measurements performed with a 25% flap on a NACA 65-210 airfoil section by Cahill [47] and on the L1T2 3-element airfoil [33] indicate that a flap gap less than $g_F/c = 0.5\% \div 1.0\%$ increases the risk of premature flap separation due to the growth of the wing wake, an early mixing with the flap boundary layer, and therefore a premature flow separation above the flap. Additionally, the off-the-surface pressure recovery is more pronounced when the the pressure imposed by the flap at the wing trailing edge is lower. Getting the flap

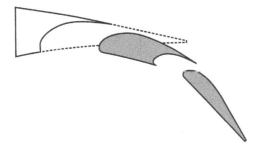

**FIGURE 5.40**   Principle of a double slotted flap.

closer reduces the dumping effect and shifts the pressure recovery more toward the wing boundary layer than into the wing wake.

## 5.2.6   DOUBLE AND TRIPLE SLOTTED FLAPS

The effect of slotted flaps, either of the simple Fowler type, can be increased by staggering the trailing edge device into several flap elements as illustrated in Figure 5.40. As the second or third flap acts on the fore flap, the same as the flap onto the wing, each slot acts as a lift amplifier. "It has long been believed that the single-slotted flap is not powerful enough to provide an acceptable landing attitude; however, the [...] Airbus models A320, A330, and A340 prove otherwise" – (Rudolph [26] p. 137). It is, therefore, not surprising that most large transport aircraft use at least double slotted flaps. On the other hand, each variable slot requires an additional mechanical system that has to enable the right deflection path together. A look at the Boeing B737 hooked track support for the triple slotted flap in Figure 5.41 shows how complicated this can get.

Figure 5.42 sketches the change in aerodynamic performance induced by the count of slots in terms of lift coefficient and lift-to-drag ratio. Every slot introduces an increase of lift coefficient, both in the linear range and in terms of maximum lift coefficient. And, there is a significant shift of the lift coefficient for approach towards lower angles of attack. The shift is more severe for the approach speed due to the scaling with the constant overspeed ratio related to the stall speed given by $C_{L,appr} = C_{L,max}/1.23^2$. This is the main reason for the belief in the failure of single slotted flaps, as it was strongly doubted that a high enough angle of attack during approach and landing could be realized. Looking into performance for take-off, every new slot causes an increase of drag coefficient and a resulting degradation of the lift-to-drag ratio. Most commonly, a multi-slotted flap is only favorable in terms of climb performance when a system with fewer slots would already be too close to stall. In multi-slotted flap systems, it is even worth to adapt the mechanics so that the slot count is reduced for partly retracted take-off settings.

An alternative to the different flap parts moving relatively together is to implement the slotted flap in a rigid assembly. The so-called fixed vane flap shown in Figure 5.43 consists of a large main flap with a smaller vane element mounted upstream and rigidly connected. Such a flap system is getting close to the aerodynamic performance

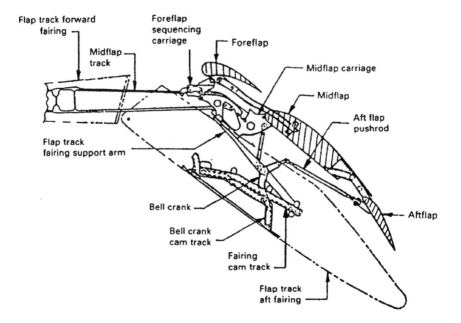

**FIGURE 5.41**   Hooked track support of the Boeing 737 flap. (From Rudolph [26], figure 1.15.)

of a double-slotted flap. But the flap support is of the same complexity as for a single slotted flap. Such a fixed vane system has been installed on the Douglas DC-9 (later MD-80/B717) and on the Airbus A400M. "For takeoff, it is generally desirable to have the vane sealed against the upper cove panel or spoilers because in this setup only the second slot is open and takeoff L/D is improved" – (Rudolph [26] p. 15). If properly designed, the vane element is completely retracted into the wing shroud in take-off position and only the gap between vane and flap is open and acting. The vane element is then exposed to the flow only in landing setting.

## 5.3   VORTEX GENERATING DEVICES

In longitudinal vortices, the flow circulates around the vortex core at a certain rate. The vorticity is defined as the local rotation of the fluid

$$\boldsymbol{\omega} = \nabla \times \mathbf{u} \tag{5.9}$$

and; therefore, a vector field. The circulation is the boundary integral of a closed surface, $S$, of the velocity along the boundary

$$\Gamma = \oint_{\delta s} \mathbf{u} \cdot d\mathbf{s} \tag{5.10}$$

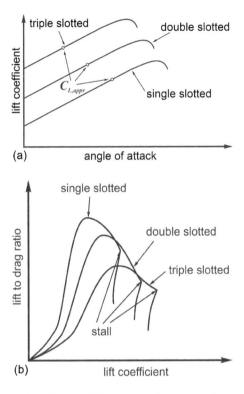

**FIGURE 5.42**  Impact of multi-slotted flaps on aerodynamic performance.

and, according to the Stoke's theorem, the integral of the vorticity orthogonal to the surface

$$\Gamma = \int_S (\nabla \times \mathbf{u}) \cdot \mathbf{n} dS = \int_S \boldsymbol{\omega} \cdot \mathbf{n} dS. \tag{5.11}$$

At the first glance, this may seem contradictory to the assumption of the rotation-free vector field in potential theory but having a circulation. But, in potential theory, the

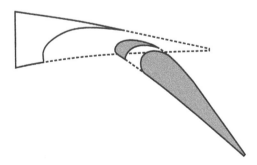

**FIGURE 5.43**  Principle of a fixed vane flap.

**FIGURE 5.44**   Principle effect of a vortex on the evolution of a wall boundary layer.

vorticity is concentrated in the singularity of the vortex core. In real viscous flow, the shear forces in the center of the vortex introduce a deviation from the potential vortex and the flow behaves there like a rigid body rotation. An approximate formulation of a viscous vortex is given by the Rankine vortex

$$\mathbf{u}(r) = \begin{cases} \Gamma r/\left(2\pi R^2\right) & \forall\, r \leq R \\ \Gamma/(2\pi r) & \forall\, r > R \end{cases}.$$ 

(5.12)

The effect of the vortex flow near a wall boundary layer is sketched in Figure 5.44. The vortex transports momentum from the outer flow to the near wall region and displaces fatigue near wall flow to the outer flow where it is accelerated by shear forces. The boundary layer flow is thereby entrained by flow with higher kinetic energy. On average, this stabilizes the boundary layer against separation.

Nevertheless, in the extent of the vortex, the flow velocity in vortex direction is on the downwash side, where the high momentum flow is pushed against the wall, higher than on the upwash side, where the low momentum flow is transported away from the wall. Therefore, the region on the upwash side of the vortex is destabilized against separation.

There are several measures to quantify the impact of the vortices to prevent flow separation. Godard and Stanislav [48] propose the relative increase of wall shear stress $\Delta\tau_w/\tau_{w,0}$ to quantify the acceleration of the boundary layer flow, with the wall shear stress given by

$$\tau_w = \mu\frac{\partial u}{\partial y}.$$ 

(5.13)

### 5.3.1   VORTEX GENERATORS

Vortex generators (VG) are devices or local adoptions of the wing that create longitudinal vortices near the boundary layer, summarized in Figure 5.45. Longitudinal vortices can be generated in two ways. The first is a local sudden change of the wing circulation. This can be most effectively achieved by a sudden change of the wing chord forming a tip-like vortex at the disruption of the wing shape. As the concentration of the longitudinal vortex is closely related to the local flow velocity, this can be achieved by a discontinuity of the leading edge (notched leading edge, dogtooth leading edge). The second method is to place a low aspect ratio disturbance at an inclination to the flow (boundary layer fence, vortex generator, Vortilon). At those

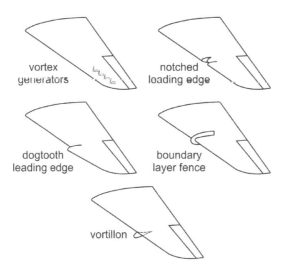

**FIGURE 5.45**   Different types of vortices generating devices. (According to [49].)

devices, a strong vortex is forming similarly to the lifting vortex of a low-aspect ratio wing.

Figure 5.46 shows several shapes in use for vortex generators. In general conclusion, the effectiveness of vortex generators is less related to the shape than to the aspect ratio of the device. However, Godard and Stanislav [48] state that "triangular actuators produce a significant improvement (+20%) compared to rectangular ones. Moreover the triangular shape is better in term[s] of drag penalty" – (Godard & Stanislav [48] p. 190).

Especially at swept wings, the local flow direction is different in cruise condition and at low speeds. In cruise condition, the stagnation point is close to the leading edge and the local flow is nearly in line-of-flight. In low-speed flight, the stagnation point moves to the lower side and the flow is highly accelerated around the leading edge. The local flow direction is therefore nearly perpendicular to the leading edge. It is therefore possible to select the flight regime in which the vortex generating device is efficient. For the generation of vortices to prevent separation, it is worth

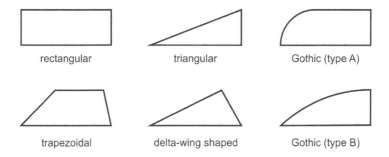

**FIGURE 5.46**   Shape types for vortex generators.

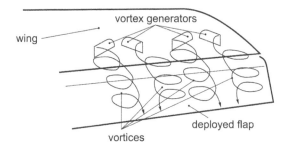

**FIGURE 5.47**   Principal sketch of an array of vortex generators on a wing. (According to [49].)

to orientate the device in the direction of flow in cruise condition. This minimizes a negative impact on cruise and the corresponding drag increment. In low-speed flight, the inclination towards the incoming flow due to the changed flow direction is responsible for the generation of the vortices when desired.

## 5.3.2   Vortex Generator Arrays

As the longitudinal vortices only act locally, it is necessary to place a grid of vortex generators if a larger area of the wing has to be influenced, as illustrated in Figure 5.47. In such grids, the vortex generators can be oriented to create co-rotating or counter-rotating vortices as shown in Figure 5.48 and Figure 5.49, respectively. Co-rotating longitudinal vortices are more smoothly influencing the flow and should be placed regularly. Counter-rotating longitudinal vortices influence the area that is called "common flow down," where both stabilizing downwash sides stabilize the flow. In the "common flow up" area, the low momentum fluid is transported away from the wall. Therefore, counter-rotating arrays of vortex generators are placed at an odd-even distribution with the vortex generators being closer together on the common flow down-side.

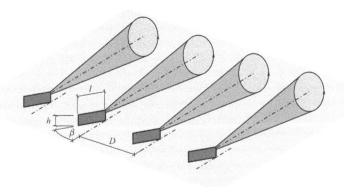

**FIGURE 5.48**   Arrangement of vortex generators inducing co-rotating longitudinal vortices.

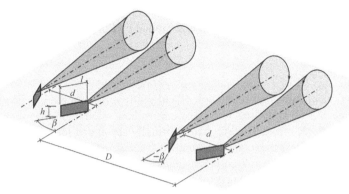

**FIGURE 5.49** Arrangement of vortex generators inducing counter-rotating longitudinal vortices.

The main geometric parameters of a vortex generator are its height $h$, its length $l$, and its inclination towards the flow direction $\beta$. Within a grid of co-rotating vortex generators, the separation distance $D$ between the vortex generators describes also the distance between the vortex cores. In counter-rotating arrangement, the separation distance $D$ describes the spacing of the vortex pairs, while the distance $d$ describes the spacing of the vortex generator pair for the common flow down region. Since the size of the vortex generator should respect the thickness of the boundary layer, they are often specified with relative values, which are the height in relation to the boundary layer thickness $h/\delta$, the aspect ratio $l/h$, and the relative spacing $D/h$ and $d/h$. Godard and Stanislav give optimal ranges for vortex generators listed in Table 5.1. "The co-rotating configuration is less effective than the counter-rotating one. The average difference is about 100%" – (Godard & Stanislav [48] p. 190). However, on swept wings, the co-rotating installation is to be preferred, as otherwise one of a pair of vortex generators has a detrimental orientation other than the desired flight conditions.

Due to the small size and corresponding low local Reynolds number, the cross-section is less important than the size and the aspect ratio. Therefore, most vortex generators are simply flat plates. The small height relative to the boundary layer thickness may surprise. However, Lin [50] states that "optimally placed $h/\delta \sim 0.2$ vane-type VGs performed just as well as a $\delta$-scale VG that has a device drag an

**TABLE 5.1**
**Optimal Geometric Parameters for Vortex Generator Arrays**

|  | $h/\delta$ | $l/h$ | $D/h$ | $d/h$ | $\beta$ | $\Delta\tau_w/\tau_{w,0}$ |
|---|---|---|---|---|---|---|
| Counter-rotating | 0.37 | 2 | 6 | 2.5 | 18° | 110 ÷ 200% |
| Co-rotating | 0.37 | 2 | 6 |  | 18° | 55 ÷ 105% |

*Source:* According to Godard & Stanislav [48] p. 188.

order-of-magnitude higher" – (Lin [50] p. 4). A comprehensive summary on experimental data on different types of VGs is compiled by Lin [51].

### 5.3.3 Slat Horn

One of the critical areas for separation of a high-lift wing is in the junction towards the fuselage. As for a normal wing, the corner flow is additionally weakened by the boundary layer of the fuselage. This is usually accounted for by an adaption of the wing twist towards the root. With high-lift systems, additionally, the slat is ended before the root and deflected perpendicular to the wing leading edge. Due to the local change in circulation, a vortex is forming at the wing root and a second one at the slat tip. Together with the horse-shoe vortex of the wing root, a system of three longitudinal vortices is formed. The paths of these vortices along the wing surface are strongly depending on the relative strengths of the vortices.

Figure 5.50(a) shows the vortex system at a normal slat side edge. The vortex generated at the wing fuselage junction transports due to its directional sense parts of the weak fuselage boundary layer towards the wing. Increasing the strength of the vortex originating from the slat side edge by a so-called "slat horn" in Figure 5.50(b), pushes the vortex originating from the junction of fuselage and wing away from the wing surface. The weakening of the wing boundary layer flow by the fuselage boundary layer is thereby prevented. Haines and Young [52] show the impact of the slat horn size on the maximum lift for two different engine sizes. Figure 5.51 reproduces their reported impact of different slat horn sizes on the lift coefficient depending on the angle of attack. The effect of the slat horn can be seen both in the maximum lift coefficient and the angle of attack of stall onset. They "showed that changing from the small to the large horn improved $C_{L,max}$ by about 0.32 in the take-off configuration, 0.1 in the approach case" – (Haines & Young [52] p. 65). The situation in the landing configuration needs some explanation as well. Here, "the region near the nacelle dictates the stall" – (Haines & Young [52] p. 65) – and the size of the slat horn is no longer as pronounced as the wing junction area is not the limiting area.

### 5.3.4 Nacelle Strakes

The impact of large nacelles in an underwing-mounted arrangement on the maximum lift behavior is well described by Chambers [53]: "At the high angles of attack required at low airspeeds, vortices are shed from the fan cowl. For engine installations where the nacelle is located further below the wing, such as JT9D installations on the 747, these vortices pass underneath the wing. For more close-coupled nacelle configurations, these vortices flow over the top of the wing and interact with the wing flow field. The effect of these vortices is generally favorable as long as they remain intact. Unfortunately the wing, at high angle of attack, will impose large adverse pressure fields on these vortices as they flow rearward along the wing surface [...]. The vortices burst when encountering the adverse pressures, resulting in boundary-layer separation on the wing and an associated loss in maximum lift" – (Chambers [53] p. 5).

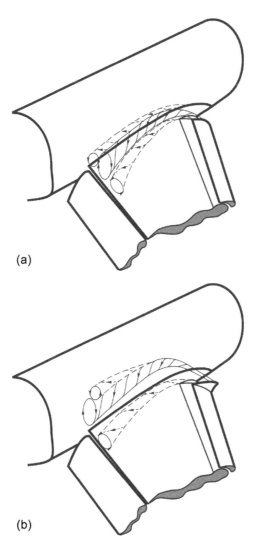

(a)

(b)

**FIGURE 5.50** Vortex system at the inboard slat side edge. (a) Normal (small) slat edge; (b) large slat horn. (Reproduction of Haines & Young [52], figure 3.90(b).)

Obert [54] describes in detail the situation that is even worsened by interrupting the leading edge to enable a pylon for a close coupled engine at the example of wind tunnel tests during the Fokker F29 aircraft project. Figure 5.52 shows the development of the wing separation with and without the cut-out of the leading-edge device by indicating the development of separation lines together with the corresponding angles of attack. Without the cut-back, a stable wing root separation develops at $\alpha = 20.5°$ and the rest of the wing stays stable up to an angle of attack of $\alpha = 26°$. Due to the cut-out of the leading-edge device, the "protection" of the wing boundary layer is missing. The wing starts to stall at $\alpha = 19°$ with the separation spreading out

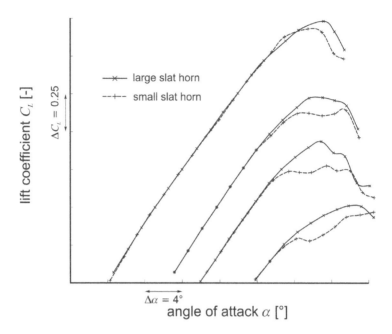

**FIGURE 5.51** Impact of slat horn size on maximum lift coefficient. (According to Haines & Young [52], figure 3.90(c).)

with increasing angle of attack. The lift behavior shows that the earlier separation is responsible for an approximate drop in maximum lift of 7.5 ÷ 10%.

Figure 5.53 shows the development of the free nacelle vortices in this area. In an ideal case, the inner nacelle vortex strikes above the wing area prone to separation and brings energized air towards the wing boundary layer. However, the nacelle vortex is resulting from a free separation, therefore not very concentrated, and most

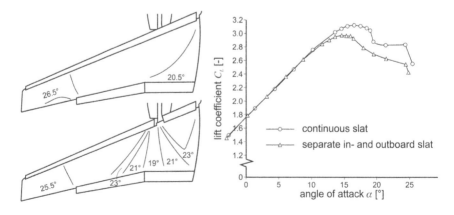

**FIGURE 5.52** Premature stall development downstream of the engine due to interruption of the leading edge device. (According to Obert [54], figure 51.)

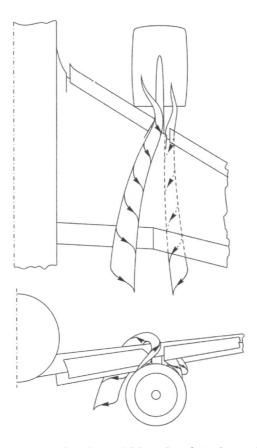

**FIGURE 5.53** Development of vortices at high angles of attack at a close-coupled under wing mounted engine nacelle. (According to Haines & Young [52], figure 3.88(d).)

unlikely at the right position. A slight outboard movement of the nacelle vortex results in the destabilizing up-wash area being just at the portion of the wing unprotected by a leading-edge device causing an additional risk for separation.

To prevent this separation, Douglas Aircraft Cooperation developed the device that is today called "nacelle strake" [55] during development of the DC-10 aircraft. As shown in Figure 5.54, reproduced from the Douglas patent [55], the idea is to use large vortex generators placed onto the nacelle to generate vortices that stabilize the boundary layer downstream of the engine. "These strakes were subjected to quite high local angles of attack due to the extreme crossflow around the nacelles at airplane angles of attack approaching maximum lift. The low aspect ratio planform of the strake promoted a strong leading edge vortex from each strake that went directly over the wing just behind the nacelle. These twin vortices energized the wing airflow sufficiently to eliminate the premature stalling and allow the wing high lift system to achieve its full maximum lift capability" – (Schaufele [56] p. 17).

Today, nacelle strakes can be found on nearly every large transport aircraft with underwing mounted engines. Schaufele [56] pleasurably writes: "In later years, we

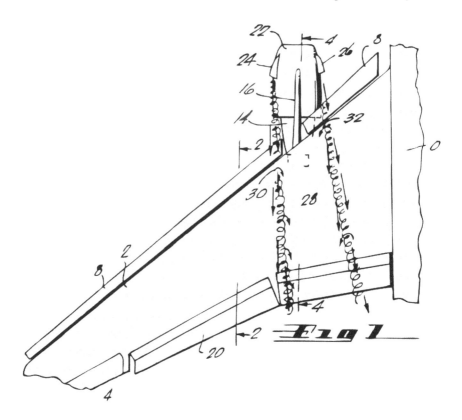

**FIGURE 5.54** Longitudinal vortices generated by nacelle strakes to prevent early wing stall downstream on a large underwing mounted engine. (from Kerker & Wells [55], figure 1.)

at Douglas were pleased to note that nacelle strakes became a standard feature of Boeing and Airbus transports" – (Schaufele [56] p. 17).

In variation to the Douglas invention, for most aircraft, a single inboard nacelle strake is sufficient. In modern understanding, the nacelle strake acts slightly differently than described before. In fact, the strake enables the positioning of the nacelle separation vortex to a favorable position. And, the vortex generator enables a more concentrated and, therefore, more effective longitudinal vortex for the re-energization of the wing boundary layer.

## NOTES

1. Weyl [9] used the term "unassisted" in the same sense.
2. The split flap with the deflection of $\delta_F = 90°$ has further been also renowned as Schrenk-flap, see [8] p. 190.
3. Woodward and Lean refer to a contribution of Smith within the AGARD CP 102 conference proceedings. Nevertheless, the publication of Smith in the Journal of Aircraft [26] of 1975 is the more recognized one.

# REFERENCES

[1] Glauert H (1927) Theoretical Relationships for an Aerofoil with Hinged Flap, ARC R&M 1095.

[2] Schlichting H, Truckenbrodt EA (2001) Aerodynamik des Flugzeuges, vol. **2**, 3rd edition, Springer, Berlin.

[3] Nayler, Stedman, Stern (1914) Experiments on an Aerofoil Having a Hinged Rear Portion, British R&M No. 110.

[4] Weyl AR (1945) High-Lift Devices and Tailless Airplanes, Aircraft Engineering, **17**(10), pp. 292–297.

[5] Spearman ML (1948) Wind-Tunnel Investigation of an NACA 0009 Airfoil with 0.25- and 0.50-Airfoil-Chord Plain Flaps Tested Independently and in Combination, NACA TN 1517.

[6] Wenzinger CJ, Rogallo FM (1939) Résumé of Air-Load Data on Slats and Flaps, NACA TN 690.

[7] Fiecke D (1956) Die Bestimmung der Flugzeugpolaren für Entwurfszwecke, DVL report no. 15, Westdeutscher Verlag/Köln.

[8] Irving HB (1935) Wing Brake Flaps, Aircraft Engineering, August 1935, pp. 189–197.

[9] Royer FE (1919) Perfectionnement aux Surfaces Portantes d'aéroplanes, Patent FR 495.551.

[10] Wright O, Jacobs JMH (1924) Airplane, patent US 1,504,663.

[11] Schrenk O, Gruschwitz E (1932) Über eine einfache Möglichkeit zur Auftriebserhöhung von Tragflügeln, Zeitschrift für Flugtechnik und Motorluftschiffahrt **23**(20), pp. 597–601. Translation to English: A Simple Method for Increasing the Lift of Airplane Wings by Means of Flaps, NACA TM 714 (1933).

[12] Liebeck RH (1978) Design of Subsonic Airfoils for High Lift, Journal of Aircraft **15**(9), pp. 547–561.

[13] Zaparka EF (1933) Aircraft and Control Thereof, Patent US 1,893,065, Reissued as Patent US Re 19,412 (1935).

[14] Henne PA (1990) Innovation with Computational Aerodynamics: The Divergent Trailing Edge Airfoil. In: Applied Computational Aerodynamics (Ed. PA Henne), Progress in Astronautics and Aeronautics, Vol. **125**, AIAA, Washington, DC.

[15] Henne PA, Gregg RD III. (1989) Divergent Trailing Edge Airfoil, Patent US 4,858,852.

[16] Richter K, Rosemann H (2006) Numerical Investigation on the Aerodynamic Effect of Mini-TEDs on the AWIATOR Aircraft at Cruise Conditions, 25th International Congress of the Aeronautical Sciences (ICAS), Paper ICAS 2006-3.9.3.

[17] Bölkow L (1940) Tragflügel mit Mitteln zur Veränderung der Profileigenschaften, German Patent DE 694,916.

[18] Reckzeh D (2003) Aerodynamic Design of the High-Lift-Wing for a Megaliner Aircraft, Aerospace Science and Technology **7**, pp. 107–119.

[19] Strüber H (2014) The Aerodynamic Design of the A350 XWB-900 High Lift System, 29th Congress of the International Council of the Aeronautical Sciences (ICAS), Paper ICAS 2014-0298.

[20] Kintscher M, Wiedemann M, Monner HP, Heintze O, Kühn T (2011) Design of a Smart Leading Edge Device for Low Speed Wind Tunnel Tests in The European Project SADE, International Journal of Structural Integrity **2**(4) pp. 383–405.

[21] Pierce D (1973) Fluid Dynamic Lift Generating or Control Force Generating Structures, US Patent 3,716,209.

[22] Kintscher M, Geier S, Monner HP, Wiedemann M (2014) Investigation of Multi-Material Laminates for Smart Droop Nose Devices, 29th Congress of the International Council of the Aeronautical Sciences, ICAS 2014, Paper 489.

[23] Achleitner J, Rohde-Brandenburger K, Rogalla von Bieberstein P, Sturm F, Hornung M (2019) Aerodynamic Design of a Morphing Wing Sailplane, AIAA Aviation 2019 Forum, AIAA Paper 2019–2816.

[24] Vasista S, De Gaspari A, Ricci S, Riemenschneider J, Monner HP, Van de Kamp B (2016) Compliant Structures-Based Wing and Wingtip Morphing Devices, Aircraft Engineering and Aerospace Technology: An International Journal, **88**(2) pp. 311–330.

[25] Krüger W (1943) Über eine neue Möglichkeit der Steigerung des Höchstauftriebes von Hochgeschwindigkeitsprofilen, AVA-Bericht 43/W/64.

[26] Rudolph PKC (1993) High-Lift Systems on Commercial Subsonic Airliners, NASA CR 4746.

[27] James VL, Hill EG (1973) Leading Edge Flap, US Patent 3,910,530.

[28] Smith AMO (1975) High-Lift Aerodynamics, Journal of Aircraft **12**(6), pp. 501–530.

[29] Liebeck RH (1973) Study of Slat-Airfoil Combinations using Computer Graphics, Journal of Aircraft **10**(4), pp. 254–256.

[30] Betz A (1922) Die Wirkungsweise von unterteilten Flügelprofilen. Berichte und Abhandig. der Wissenschaftlichen Gesellschaft für Luftfahrt, Heft 6. Translation to English: Theory of the Slotted Wing, NACA TN 100 (1922).

[31] Hoerner FS, Borst HV (1985) Fluid Dynamic Lift, 2nd ed., Hoerner Fluid Dynamics, Bakersfield, CA.

[32] Wild J (2013) Mach and Reynolds Number Dependencies of the Stall Behavior of High-Lift Wing-Sections, Journal of Aircraft **50**(4), pp. 1202–1216.

[33] Woodward DS, Lean DE (1993) Where is High-Lift Today? – A Review of Past UK Research Programs, AGARD CP 515, pp. 1–1 – 1–45.

[34] Cole JB, Weiland RH (1970) Aircraft Wing Variable Camber Leading Edge Flap, US Patent 3,504,870.

[35] Boeing Commercial Airplane Group (1999) High Reynolds Number Hybrid Laminar Flow Control (HLFC) Flight Experiment, Part II. Aerodynamic Design, NASA CR 1999-209324.

[36] Iannelli P, Wild J, Minervino M, Strüber H, Moens F, Vervliet A (2013) Design of a High-Lift System for a Laminar Wing, 5th European Conference for Aeronautics and Space Sciences (EUCASS), Munich, Germany.

[37] Franke DM, Wild J (2016), Notes in Numerical Fluid Mechanics and Multidisciplinary Design **132**, Springer Verlag, Cham, pp. 17–27.

[38] Strüber H, Wild J (2014) Aerodynamic Design of a High-Lift System Compatible with a Natural Laminar Flow Wing within the DeSiReH Project, 29th Congress of the Council of the Aeronautical Sciences ICAS 2014, St. Petersburg, Russia, Paper ICAS 2014-0300.

[39] Wild (2016) Recent Research Topics in High-Lift Aerodynamics, CEAS Aeronautical Journal **7**(3), pp. 345–355.

[40] Handley Page F (1922) Means for Balancing and Regulating the Lift of Aircraft, US Patent 1,422,614.

[41] Wenzinger CJ, Delano JB (1938) Pressure Distribution over an N.A.C.A. 23012 Airfoil with a Slotted and a Plain Flap, NACA TR 633.

[42] Wenzinger CJ, Harris TA (1939) Wind Tunnel Investigation of an N.A.C.A. 23012 Airfoil with Various Arrangements of Slotted Flaps, NACA TR 664.

[43] Fowler HD (1921) Variable Area Wing, US Patent 1,392,005.

[44] Fowler HD (1928) Aerofoil, US Patent 1,670,852.

[45] Fowler HD (1936) The Fowler Flap Wing – The Originator's Own Description of the Theory and Uses of this Interesting Device, Aircraft Engineering **8**(9), pp. 247–249.

[46] Platt RC (1936) Aerodynamic Characteristics of a Wing with Fowler Flaps including Flap Loads, Downwash, and Calculated Effect on Take-off, NACA TR 534.

[47] Cahill JF (1947) Two-Dimensional Wind-Tunnel Investigation of Four Types of High-Lift Flap on an NACA 65-210 Airfoil Section, NACA TN-1191.

[48] Godard G, Stanislav M (2006) Control of a Decelerating Boundary Layer. Part 1: Optimization of Passive Vortex Generators, Aerospace Science and Technology **10**, pp. 181–191.

[49] www.aerospaceweb.org (access 2017) http://www.aerospaceweb.org/question/aerodynamics/q0228.shtml.

[50] Lin JC (1999) Control of Turbulent Boundary-Layer Separation using Micro-Vortex Generators, 30th Fluid Dynamics Conference, AIAA Paper 1999–3404.

[51] Lin JC (2002) Review of Research on Low-Profile Vortex Generators to Control Boundary-Layer Separation, Progress on Aerospace Sciences **38**, pp. 389–420.

[52] Haines AB, Young AD (1994) Scale Effects on Aircraft and Weapon Aerodynamics, AGARD AG 323.

[53] Chambers JR (1986) High-Angle-of-Attack Aerodynamics: Lessons Learned, 4th Applied Aerodynamics Conference, AIAA Paper 1986–1774.

[54] Obert A (1993) Forty Years of High-Lift R&D – An Aircraft Manufacturer's Experience, AGARD CP 515, pp. 27–1 – 27–28.

[55] Kerker A, Wells OD (1973) Lift Vanes, US Patent 3,744,745.

[56] Schaufele RD (1999) Applied Aerodynamics at the Douglas Aircraft Company – A Historical Perspective, 37th Aerospace Sciences Meeting and Exhibit, AIAA Paper 1999-0118.

# 6 Active High-Lift Systems

## NOMENCLATURE

| Symbol | Units | Description | Symbol | Units | Description |
|---|---|---|---|---|---|
| $A_{ref}$ | $m^2$ | Reference wing area | $Re$ | – | Reynolds number |
| $A_{slot}$ | $m^2$ | Slot/exhaust area | $t$ | $s$ | Time |
| $AFM_1$ | – | First aerodynamic figure of merit | $T$ | $s$ | Time period |
| $b$ | $m$ | Wing span | $T$ | $N$ | Thrust force |
| $c$ | $m$ | Airfoil chord | $u$ | $m/s$ | Local velocity |
| $c_F$ | $m$ | Flap chord | $U$ | $m/s$ | Flow velocity |
| $c_L$ | – | Section lift coefficient | $w$ | $m$ | Slot width |
| $c_p$ | – | Pressure coefficient | $x, y, z$ | $m$ | Coordinate |
| $c_P$ | – | Power coefficient | $\alpha$ | ° | Angle of attack |
| $c_q$ | – | Mass flow coefficient | $\beta$ | ° | Skew angle |
| $c_\mu$ | – | Momentum coefficient | $\Gamma$ | $m^2/s$ | Circulation |
| $C_D$ | – | Drag coefficient | $\delta$ | $m$ | Boundary layer thickness |
| $C_L$ | – | Lift coefficient | $\delta_F$ | ° | Flap deflection angle |
| $C_{L,max}$ | – | Maximum lift coefficient | $\delta_j$ | ° | Jet inclination/deflection angle |
| $C_m$ | – | Pitching moment coefficient | $\delta_2$ | $m$ | Momentum loss thickness |
| $C_T$ | – | Thrust coefficient | $\eta$ | – | Efficiency |
| $d$ | $m$ | Distance | $\lambda$ | – | Local velocity ratio |
| $D$ | $m$ | Separation distance | $\Lambda$ | – | Velocity ratio to free stream |
| $D_j$ | $m$ | Engine jet diameter | $\Lambda$ | – | Wing aspect ratio |
| $DC$ | – | Duty cycle | $\rho$ | $kg/m^3$ | Density |
| $f$ | $1/s$ | Frequency | $\tau_w$ | $N/m^2$ | Wall shear stress |
| $F$ | $N$ | Force | $\Delta$ | | Difference operator |
| $F^+$ | – | Dimensionless frequency | $AFC$ | | Active flow control |
| $FM$ | – | Figure of merit | $corr$ | | Corrected value |
| $l$ | $m$ | Length | $e$ | | Boundary layer edge |
| $L$ | $N$ | Lift force | $j$ | | Jet |
| $\dot{m}$ | $kg/s$ | Mass flow | $max$ | | Maximum value |
| $M$ | – | Mach number | $n$ | | Normal |
| $P$ | $W$ | Power | $0$ | | Original, not active |
| $R$ | $m$ | Flap impingement height | $\infty$ | | Free stream |
| $\Delta R_{Vx}$ | $kg \cdot m/s^2$ | Change in momentum | | | |

Active high-lift resembles all means to increase the lift coefficient by changing the energy content of the fluid. As the active change of the energy is closely coupled to the need of controlling it, the active high-lift systems are also known as Active Flow Control (AFC) technology.

DOI: 10.1201/9781003220459-6

AFC is collecting methods to remove low energy fluid from the flow field (suction) as well as those for increasing the kinetic energy of the flow (blowing). In any case, an external source of energy has to be provided. This energy source can be electric or pneumatic. In some cases described, the energy source is directly the aircraft engine by making use of the engine jet.

In relation to high-lift aerodynamics, there is a distinction between the prevention of flow separation by stabilizing the boundary layer, called boundary layer control, and the direct increase of circulation called circulation control.

In order to quantify the effort of AFC, it is important to relate the effort to the flow properties. The momentum coefficient

$$c_\mu = \frac{\dot{m} \cdot u_j}{\frac{1}{2}\rho_\infty U_\infty^2 A_{ref}} = 2 \cdot \frac{\rho_j}{\rho_\infty} \left(\frac{u_j}{U_\infty}\right)^2 \frac{A_{slot}}{A_{ref}} \tag{6.1}$$

according to Poisson-Quinton [1] relates the momentum introduced or removed to the momentum of the free stream flow. In case of an under-expanded jet, additionally, the pressure difference between the jet and the external flow has to be accounted for [2]

$$c_\mu = \frac{\dot{m} \cdot u_j + (p_j - p) \cdot A_{slot}}{\frac{1}{2}\rho_\infty U_\infty^2 A_{ref}}. \tag{6.2}$$

The mass flow coefficient

$$c_q = \frac{\dot{m}}{\rho_\infty U_\infty A_{ref}} = \frac{\rho_j}{\rho_\infty} \left(\frac{u_j}{U_\infty}\right) \frac{A_{slot}}{A_{ref}} \tag{6.3}$$

non-dimensionalizes the mass flow added to or removed from the surrounding flow.

Besides this, especially for blowing, it is important to know the ratio of the jet velocity to the velocity in the flow field. The velocity ratio is either defined globally to the free stream velocity

$$\Lambda = \frac{u_j}{U_\infty} \tag{6.4}$$

or locally with respect to the flow local velocity at the outer edge of the boundary layer

$$\lambda = \frac{u_j}{U_e}. \tag{6.5}$$

In incompressible theory, the local velocity ratio is more easily expressed by the pressure coefficient

$$\lambda = \frac{1}{\sqrt{1-c_p}} \left(\frac{u_j}{U_\infty}\right), \tag{6.6}$$

as this value is easier to be measured or identified.

The effectiveness of AFC for lift generation can be measured as the ratio of the lift improvement to the introduced momentum

$$\eta = \frac{\Delta C_L}{c_\mu}. \tag{6.7}$$

Since there is an energy needed to influence the flow, it is necessary to respect this for the evaluation of the efficiency of AFC. Greenblatt and Wygnanski [3] defined a figure of merit including the power input. The required power input can be non-dimensionalized as

$$c_P = \frac{P_{AFC}}{\frac{1}{2}\rho_\infty A_{ref} U_\infty^3}. \tag{6.8}$$

The aerodynamic figure of merit is then defined as

$$FM = \left(\frac{C_L}{C_D + c_P}\right) \Big/ \left(\frac{C_L}{C_D}\right). \tag{6.9}$$

Note that the lift coefficient is not eliminated, as the optimum aerodynamic efficiency may be obtained at different flow conditions and different lift levels. Based on this approach, the efficiency of an AFC system to provide an air flow can be defined as

$$\eta_{AFC} = \frac{c_\mu}{c_P}. \tag{6.10}$$

Another parameter to classify the effectivity of AFC especially regarding high-lift system is the so-called lift gain factor (LGF) [2]

$$LGF = \frac{C_{L,\max,AFC} - C_{L,\max,ref}}{c_\mu}, \tag{6.11}$$

which relates the improvement of the maximum lift coefficient to the introduced momentum coefficient. It is a measure of the capability of the corresponding AFC technology for lift generation.

## 6.1   BOUNDARY LAYER CONTROL (BLC)

As discussed in Chapter 1, according to Prandtl [4], flow separation occurs due to the momentum loss by wall friction in the boundary layer attached to the wall. Preventing a separation is, therefore, possible by reducing or eliminating this momentum loss. There are, in principle, two possibilities to achieve this. The first is to subtract the low energy flow of the near wall boundary layer from the surrounding flow field by suction. This was first demonstrated already by Prandtl, as discussed in Chapter 1.

The second is to increase the energy of the wall boundary layer. The latter is mostly achieved by blowing fluid with higher energy into the fluid, but also other means of acceleration are available.

### 6.1.1 BOUNDARY LAYER SUCTION

Weyl writes already in 1945: "The idea of sucking away the boundary layer is fairly old and based on the scientific aspect of fluid flow; hence it is not a discovery or invention in the common sense. The eminent aerodynamicist Professor Ludwig Prandtl of Goettingen suggested and applied this method to demonstrate that his (new) conception of the boundary layer was not a mere fiction of useless speculation but a real fact" – (Weyl [5] p. 325). The first patent on boundary layer suction to prevent flow separation was issued in 1928 to Betz and Ackeret [6]. A sound summary of applications at the mid-1940s has been compiled by Maillart [7].

Figure 6.1 shows the effect of discrete suction on the velocity profile of a weakened boundary layer. The weakness of the incoming boundary layer is indicated by the inflection point in the velocity profile. In Section 2.3, this was identified as a prerequisite for a risk of separation.

The weak lower part of the boundary layer is removed by sucking the air through the slot of the surrounding flow field. The energy source to be provided for this is some kind of suction pump or vacuum pump. The mass flow can be controlled by the suction pressure and the slot width.

Past the suction slot, the boundary layer shows a stable convex velocity profile. As the weak airflow has been removed, also the characteristic dimensions of the boundary layer – boundary layer thickness, displacement thickness, and momentum loss thickness – are reduced. Therefore, not only the separation onset is reduced, but also the remaining boundary layer is more stable against a pressure rise.

As suction is never directionally, the slot direction must not be in the direction of air flow as depicted in Figure 6.1. Suction can be applied by any kind of hole. Experiments showed that forward facing suction slots as depicted provide a higher flexibility of the applied suction rate, especially in compressible flow [8].

Figure 6.2 shows the effect of suction through a slot at the leading edge of a NACA 63₁-012 airfoil with a camber flap deflected by 40° on the maximum lift

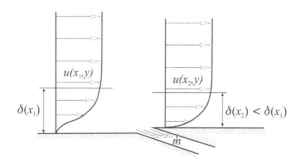

**FIGURE 6.1**    Effect of discrete suction on the velocity profile of the wall boundary layer.

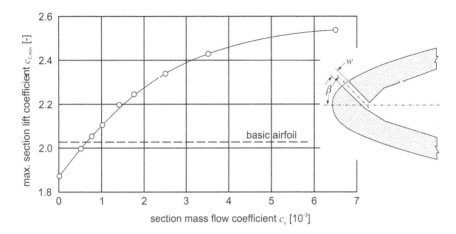

**FIGURE 6.2** Variation of maximum lift coefficient with flow coefficient for a NACA 63₁-012 airfoil with the flap deflected 40° (According to McCullough & Gault [9], figures 2 & 7.)

coefficient [9]. First, it is seen that the slot itself disturbs the flow, and the maximum lift coefficient is lower than for the airfoil without slot. An increase of maximum lift is achieved with a reasonable mass flow coefficient of $c_q \approx 0.8\%$. At high suction levels, the gain in maximum lift coefficient tends to level out. This is expected, as suction cannot provide more effect than removing the complete boundary layer, and the stall behavior will be the same above a certain level of suction.

Another extreme case of discrete suction is the application in the rear part of the adverse pressure recovery of a thick laminar airfoil. Figure 6.3 reproduces the airfoil

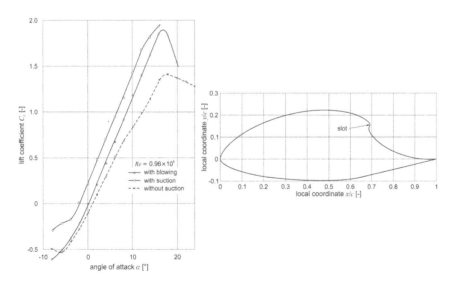

**FIGURE 6.3** Variation of lift with incidence, with and without suction, and with blowing. (According to Glauert et al. [10], figures 1 & 6.)

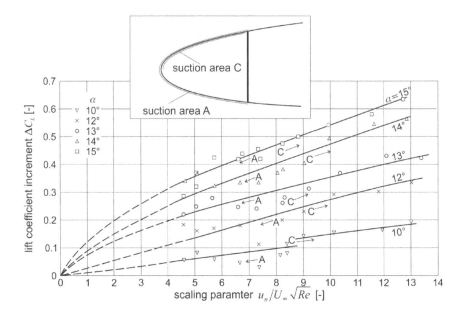

**FIGURE 6.4** Lift increment due to suction: variation with normal velocity through surface. (According to Pankhurst et al. [11], figure 4.)

and the lift coefficient characteristics of such an airfoil as measured by Glauert et al. [10]. As the pressure rise should be done on a very short distance, active suction has been applied to prevent early separation. These experiments not only showed the prevention of separation for the desired range of incidences but also a significant increase in maximum lift coefficient. As a side result of the study, also blowing was tested, showing a similar effect at a slightly lower effectiveness.

Suction must not necessarily be applied by discrete slots. Also, distributed suction can achieve a similar effect. Pankhurst et al. [11] performed wind tunnel tests with an airfoil with a porous leading edge. Figure 6.4 resumes their results on the lift increment achieved with a porous leading edge with two different extents of the suction area. They concluded that mainly the upper surface suction (area C) is responsible for increasing lift coefficients. Due to the reduced suction area, higher suction velocities can be obtained than for a full porous leading edge (area A). "Results for a range of tunnel speeds at a given incidence confirmed that, over the limited speed range of the tests, $\Delta C_L$ does not vary greatly with Reynolds number provided $u_n/U_\infty\sqrt{Re}$ (or $c_q\sqrt{Re}$) is maintained constant" – (Pankhurst et al. [11] p. 4).

Flight tests with distributed suction have been conducted by the Mississippi State University in the 1950s and 1960s. "The concept of distributed suction boundary-layer control for the prevention of separation had been developed and used to increase the maximum lift coefficients of gliders and powered aircraft" – (Bridges [12] p. 1650). The research included tests on a Schweitzer TG-3A sailplane and a Piper L-21 aircraft with suction through a porous wing surface. "The maximum lift coefficient that could be obtained from the glider with no boundary-layer control applied was 1.38 at an airspeed of 38.5 mph [33.5 kts]. [...] With [...] application of porosity, the maximum lift coefficient of the glider was increased to 1.61 at an airspeed of 35.5 mph [30.8 kts]" – (Bridges [12] p. 1644). "With the suction boundary-layer control [on the Piper L-21] and the modifications to the flap and wing leading-edge geometries, the maximum lift coefficient for full flaps and full power was increased from 3.36 to 5.62, and the maximum lift coefficient for full flaps and idle power was increased from 2.23 to 4.40" – (Bridges [12] p. 1648).

### 6.1.2  TANGENTIAL BLOWING

The opposite type of boundary layer control is adding kinetic energy to the boundary layer. This can be achieved by blowing high-energy air into the fluid along the wall. Figure 6.5 sketches the principal effect of blowing tangentially along the wall on the development of an unstable boundary layer. The region near the wall is filled up with a high energy velocity profile eliminating at this position the risk of flow separation. In comparison to boundary layer suction, the boundary layer thickness is increased. Depending on the amount of blowing, a shear layer above the boundary layer develops and the velocity deficit of the original boundary layer still exists at a slight displacement of the wall. As discussed in the context of slotted airfoils (Section 5.2), a free shear layer is more stable against a pressure rise than a boundary layer, and the low energy there is no longer directly responsible for a flow separation.

Tangential blowing has first been used to try to prevent the separation on a highly deflected plain flap to allow for higher deflection angles. Figure 6.6 reproduces wind tunnel measurements reported by Schwier [14]. The effect of blowing is highly non-linear with respect to the introduced mass flow. A minimum flow rate is needed to see a positive effect on the lift coefficient. Too less air flow disturbs the flow more than the introduced energy helps to keep the flow attached. After reaching the minimum level, a high sensitivity is seen tending to level out at higher mass flow rates.

A tangentially blown flap operating in the BLC regime was demonstrated on a large jet-driven airplane Boeing 367-80 [15], the prototype of the famous Boeing 707. In cooperation, Boeing and NASA modified the prototype to be equipped with a BLC flap. It has been the only demonstration of BLC in the class of large passenger

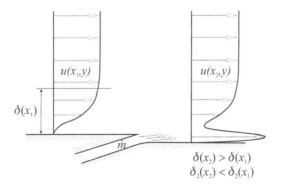

**FIGURE 6.5**  Effect of tangential blowing on the velocity profile of the wall boundary layer. (According to [13].)

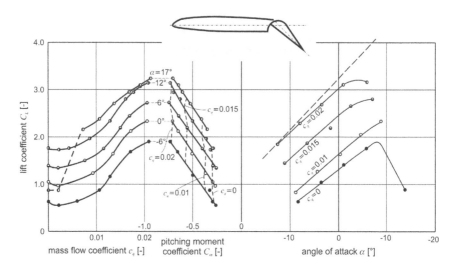

**FIGURE 6.6** Impact of tangential blowing at the shoulder of a plain flap. (According to Schwier [14], figure 15.)

transport jet aircraft. In contrast to the original double slotted flap, the aircraft was equipped with a continuous simply hinged plain flap. Pressurized air was exhausted above the flap and controlled by an ejector nozzle. "The modifications to the wing and ejector resulted in flow attachment through the full range of flap deflection to 85°. [...] the airplane was flown throughout its low-speed flight envelope and landed using up to 70° of flap and maximum blowing. The lowest approach speed utilized was 78 knots equivalent airspeed (EAS)" – (Gratzer & O'Donnell [15] p. 482).

Tangential blowing at the nose has been performed e.g. within the European project EUROLIFT II to delay stall at a 2-element (wing and flap) configuration and to recover the performance of a 3-element configuration with slat [16]. Lift coefficients measured are shown in Figure 6.7. The removal of a slat for this configuration resulted in a drop in lift coefficient by about $\Delta C_L \approx 0.75$. With a slight blowing with an already significant velocity ratio related to the global flow velocity of $\Lambda = 1.25$, no positive effect was achieved. In contrast, stall occurred earlier and at a lower incidence. In fact, the velocity ratio had to be increased up to $\Lambda \approx 2.5$ to recover the aerodynamic lift performance of the 2-element configuration without any blowing. At the maximum velocity ratio of $\Lambda = 5.0$, the exit velocity of the jet achieved sonic speed limiting the capabilities of this type of blowing. With a lift increase of $\Delta C_L \approx 0.42$, the removal of the slat was recovered by slightly more than half.

### 6.1.3 UNSTEADY BLOWING

The efficiency of AFC by blowing can be increased when considering blowing intermittently instead of continuously. As the effectiveness of blowing is related to the velocity ratio and the momentum coefficient, it is possible to achieve the same effect with less mass flow. Seifert et al. [17] summarized the superior efficiency of the periodic excitation over steady blowing.

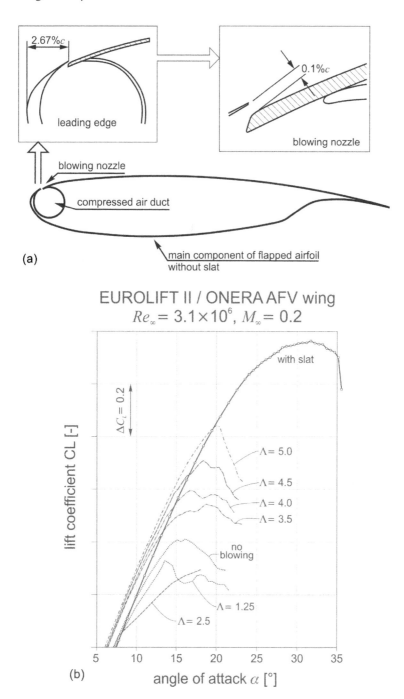

**FIGURE 6.7** Impact of tangential blowing at the leading edge to prevent separation. (According to Wild, Chan, & Rudnik [16] viewgraphs 11 & 18.)

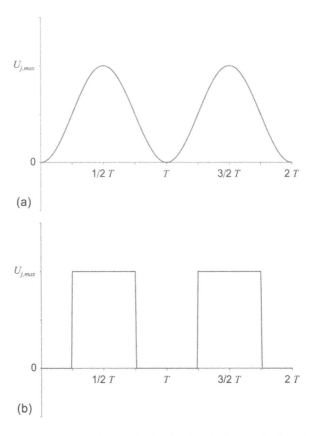

**FIGURE 6.8**  Velocity signals of unsteady blowing jets: (a) harmonic; (b) pulsed.

Commonly, two types of time signals shown in Figure 6.8 have been applied to achieve a timely varying flow actuation. The first is a harmonic oscillation of the jet flow (Figure 6.8a), the second is called pulsed blowing as it is targeted to achieve the jet exit velocity in a square signal (Figure 6.8b).

In order to compare the efficiency, it is now important to account for the unsteady jet velocity also in the evaluation of the characteristic coefficients. As the velocity amplitude is periodic, it is necessary to base the coefficients on the time averages. For the momentum coefficient, this results in

$$c_\mu = \frac{A_{slot} \dfrac{1}{T} \displaystyle\int_T \rho_j(t) u_j(t)^2 \, dt}{\rho_\infty U_\infty^2 A_{ref}} \tag{6.12}$$

and, for the mass flow coefficient, in

$$c_q = \frac{A_{slot} \dfrac{1}{T} \displaystyle\int_T \rho_j(t) u_j(t) \, dt}{\rho_\infty U_\infty A_{ref}}. \tag{6.13}$$

The unsteady blowing introduces two new characteristic parameters describing the time signal. The first is the frequency of the blowing. It is worth to relate the blowing frequency to a characteristic time scale of the controlled flow. The dimensionless frequency is defined similar to a Strouhal number by relating the frequency to the time scale the surrounding flow needs to travel along the controlled body

$$F^+ = \frac{f \cdot c}{U_\infty}, \tag{6.14}$$

where the reference length $c$ is either chosen to be the distance between the flow control actuator and the trailing edge of the corresponding airfoil [18] or the chord length of the controlled airfoil element itself [19, 20]. The second characteristic value is the so-called duty cycle $DC$ that describes the portion of the period where the flow control is acting

$$DC = \frac{t_j}{T}. \tag{6.15}$$

While, for the harmonic blowing, the duty cycle is constant $DC = 0.5$, especially for the pulsed blowing, the duty cycle is a controllable parameter. It links the average velocity to the maximum peak velocity

$$\bar{u}_j = DC \cdot u_{j,\mathrm{max}}, \tag{6.16}$$

as well as the average momentum to the peak momentum

$$\frac{1}{T}\int_T \rho_j(t)u_j(t)^2 \, dt = \overline{\left(\rho_j u_j^2\right)} = DC \cdot \rho_j u_{j,\mathrm{max}}^2, \tag{6.17}$$

and the average mass flow assuming constant jet density

$$\bar{m} = \frac{1}{T}\int_T \rho_j(t)u_j(t)\,dt = \overline{\left(\rho_j u_j\right)} = DC \cdot \rho_j u_{j,\mathrm{max}}. \tag{6.18}$$

The corresponding momentum coefficient is therefore

$$c_\mu = \frac{\overline{\left(\rho_j u_j^2\right)} \cdot A_{slot}}{\rho_\infty U_\infty^2 A_{ref}} = \frac{1}{DC}\frac{\bar{m} \cdot \bar{u}_j}{\rho_\infty U_\infty^2 A_{ref}} = \frac{\bar{m} \cdot u_{j,\mathrm{max}}}{\rho_\infty U_\infty^2 A_{ref}}. \tag{6.19}$$

"Since there is a factor $1/DC$ in the pulsed blowing definition, it can be noted that to obtain the same $c_\mu$ coefficient between steady and unsteady blowing, the mean mass flow rate $\bar{m}$ in the unsteady blowing case must be decreased by a factor $\sqrt{DC}$" – (Meunier [18] p. 2).

Meunier and Dandois [21] performed simulations comparing the efficiency of pulsed blowing in comparison to continuous blowing at the hinge of a plain flap to recover the aerodynamic performance of a slotted Fowler flap. Figure 6.9 shows their

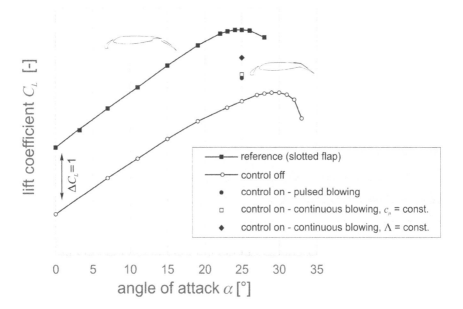

**FIGURE 6.9**  Scalability of effectivity of flow control by pulsed tangential blowing. (According to Meunier & Dandois [21], figure 14.)

reported impact of AFC on the lift coefficient vs. angle of attack. They showed that, by continuous blowing, it is possible to completely recover the performance of the Fowler type flap but at high costs of supplied pressurized air. Afterward, they simulated the pulsed blowing with $DC = 0.5$ and compared it to steady blowing once at the same momentum coefficient and once at the same velocity ratio[1]. First, it was concluded that the effect of tangential blowing is the same at the same momentum coefficient, regardless of whether the blowing is steady or pulsed. Second, at the same velocity ratio, the effect of blowing scales with the mass flow is defined by the duty cycle. This study shows that, in tangential blowing, the introduced momentum is the scaling parameter.

The effectivity of pulsed blowing for separation control is improved if the blowing direction is inclined from the surface. Petz and Nitsche [22] and Schatz et al. [23] successfully attempted to stimulate instabilities of the boundary layer to efficiently prevent separations above the flap of a multi-element airfoil wing. Later, Haucke et al. [19] showed that inclining the blowing direction at $30° \div 45°$ related to the surface achieved better efficiency.

Figure 6.10 shows the principal effect of the inclined blowing. In addition to introduction of momentum into the boundary layer, a secondary flow structure in terms of a transversal vortex appears that enhances the mixing of outer flow into the weak boundary layer. It is important to note that this additional flow structure is only beneficial if it is unsteady. Steady blowing through the inclined slot would result in a separated flow past the blowing location. Due to the pulsed blowing, when the blowing is activated, the transversal vortex is generated. When blowing is deactivated, the vortex is "released" from the blowing location and travels downstream along

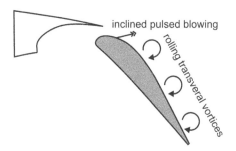

**FIGURE 6.10**  Principal flow field development by pulsed downstream blowing inclined to the surface.

the surface. Experimental investigations [19] as well as numerical studies [20] have shown that, if the frequency is high enough to have at least one vortex above the surface, the flow field shows an in average attached behavior.

Figure 6.11 shows the corresponding average pressure distribution of the DLR-F15 2-element airfoil as measured by Haucke et al. [19] with and without inclined pulsed blowing. The corresponding frequency here equals a dimensionless frequency $F^+ = 0.42$ related to flap chord, which is in line with the investigations of Seifert et al. [17] that have shown an increasing effectivity of AFC up to a value of $F^+ = 1.0$. At higher frequencies, the effectiveness of flow control is only marginally affected.

### 6.1.4  VORTEX GENERATING JETS

"Small jets blown through holes in the solid surface can generate longitudinal stream-wise vortices in a boundary layer, and these vortices act to increase cross-stream

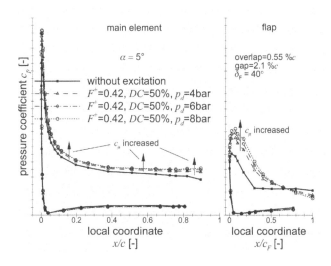

**FIGURE 6.11**  Pressure distribution with and without pulsed inclined blowing at the flap of the DLR-F15 2-element airfoil. (From Haucke et al. [19], figure 11.)

mixing of streamwise momentum" – (Johnston [24] p. 989). The effect of longitudinal vortices on the flow and their ability to delay flow separation has been discussed in the scope of vortex generators (Section 5.3). As those jets act in a similar way with a fluidic source for the vortices, they are called vortex generating jets (VGJ).

Since vortex-generating devices don't act homogenously over the span but in a discrete manner, local measures are beneficial to justify the positive effect on the boundary layer flow. As for the mechanical vortex generators, one measure is the maximum increase in wall shear stress as defined by Godard [25] according to eq. (5.13). Nevertheless, this measure is not able to account for the average effect of spanwise varying actuation. Ortmanns et al. [26] define a benefit measure for the "change in momentum due to viscous forces as a result of the acting vortex structure relative to the undisturbed two dimensional boundary layer flow" – (Ortmanns et al. [26] p. 405) – in terms of a spanwise integral of the momentum increase above the surface.

$$\Delta R_{V_x}(x,z) = \frac{\int_b \left( u(x,y,z)^2 - u_0(x,y,z)^2 \right) dy}{\int_b \left( u_0(x,y,z)^2 \right) dy} \tag{6.20}$$

Since the integral is depending on the position of the integration line, "the altitude of this line – as long as it is positioned beneath the vortices and reasonable closed to the surface – might have influence on the quantity of $\Delta R_{V_x}$, but changes in efficiency of different configurations will still be described correctly" – (Scholz et al. [27] p. 5).

In principle, there are two types of vortex-generating jet devices [28]. The first type uses thin slots perpendicular to the wall surface and inclined to the oncoming flow. The second type uses round holes that blow inclined to the wall surface and perpendicular to the flow.

Figure 6.12 shows the principal geometric arrangement of the blowing slits in counter-rotating orientation to generate longitudinal vortices similar to the vortex

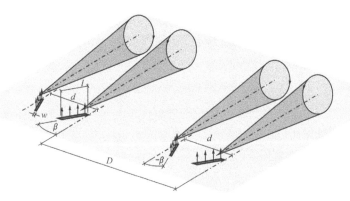

**FIGURE 6.12** Vortex generator jets by blowing through slits as source for longitudinal vortices.

## TABLE 6.1
### Optimal Geometric Parameters for Counter-Rotating Slit Vortex Generator Jets

|                    | $\Lambda$ | $w/l$ | $d/\delta$ | $d/l$ | $D/l$ | $\beta$ | $\Delta\tau_w/\tau_{w,0}$ |
|--------------------|-----------|-------|------------|-------|-------|---------|---------------------------|
| Counter-rotating   | 6         | 1/15  | 0.099      | 2.08  | 6.87  | 15°     | 55 ÷ 70%                  |

*Source:* Godard, Foucaut, & Stanislav [29] p. 396.

generators in Section 5.3.2. The air is blown out orthogonally to the surface and the slit orientation is inclined with the skew angle $\beta$ to the flow direction. Godard et al. [29] analyzed the geometric characteristics of optimal slit actuators listed in Table 6.1. The fluidic slit actuators achieve roughly half of the effectivity compared to rigid vortex generators. An additional investigation assessed pulsing of the jet flow to reduce the required effort. Their analysis showed, at a sufficiently high frequency, a comparable effectivity if the velocity ratio is similar. It has to be highlighted that these investigations have been performed with a flat plate boundary layer flow. In this case, the local flow velocity at the boundary layer edge is approximately the same as the free stream value. For applying fluidic vortex generator jets at an airfoil with a significant acceleration of the flow and an adverse pressure gradient, it is important to account for the local velocity ratio, $\lambda$, not the global velocity ratio, $\Lambda$.

While the vortex generator jets with slotted outlets have a geometrical relation to the rigid vortex generators, generating vortices by blowing through round holes has a slightly different characteristics. With this type of device, air is blown in a direction inclined to the surface and perpendicular to the incoming flow.

Godard & Stanislas [30] studied also this device in a parametric study, accounting for co-rotating and counter-rotating arrangements as shown in Figure 6.13. Their determined optimum geometrical properties are listed in Table 6.2. As only a pair of holes was investigated, this study did not account for the ratio of distances for actuator pairs in counter-rotating arrangement. "In contradiction to the passive devices [...] both configurations lead to comparable skin friction increase" – (Godard & Stanislav [30] p. 462). In conclusion, "for the counter-rotating case, the results obtained are fairly comparable between the jets and the passive actuators" – (Godard & Stanislas [30] p. 462).

A direct comparison of the two types of vortex generator jets and rigid vortex generators at 25% chord of a 12% thick, low drag type of two-dimensional aerofoil has been performed by Wallis [28]. He measured the stagnation pressure of the near wall flow close to the trailing edge. On the airfoil, rigid vanes and both types of fluidic vortex generators were installed. Figure 6.14 shows the head pressure measured in a spanwise traverse passing the wake of one co-rotating pair of devices each. As confirmed in the later studies of Godard, the round vortex generator jets ("inclined jets") perform similar to the rigid vortex generators ("metal vanes"), while the slot-type vortex generator jets ("air vanes") achieve a significantly lower effectivity.

Although the effectivity of round jet vortex generator jets is similar to rigid vortex generators, the details of the flow differ significantly. The measurements of the

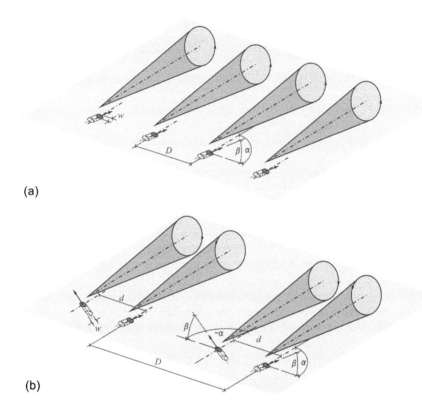

(a)

(b)

**FIGURE 6.13** Vortex generator jets by blowing through holes as source for longitudinal vortices (a) co-rotating; (b) counter-rotating.

boundary layer traverse by Wallis shown in Figure 6.15 at the most effective positions marked in Figure 6.14 show that the rigid vortex generator shows a distinct velocity maximum indicating a discrete embedded vortex, while, for the round vortex generator jet, the entrainment of the boundary layer by the outer flow is more complete and the boundary layer shows the typical shape of a turbulent boundary layer. The boundary layer profiles indicate that both the wall skin friction and the momentum loss thickness are slightly lower with the fluidic vortex generators.

**TABLE 6.2**

**Optimal Geometric Parameters for Co-Rotating and Counter-Rotating Vortex Generator Jets**

|  | $\Lambda$ | $w/\delta$ | $D/w$ | $d/w$ | $\alpha$ | $\beta$ | $\Delta\tau_w/\tau_{w,0}$ |
|---|---|---|---|---|---|---|---|
| Counter-rotating | >3.1 | 0.036 |  | 15 | 45° ÷ 90° | 45° | 200% |
| Co-rotating | 4.7 | 0.036 | 6 |  | 45° ÷ 90° | 45° | 220% |

*Source:* Godard & Stanislav [30] p. 462.

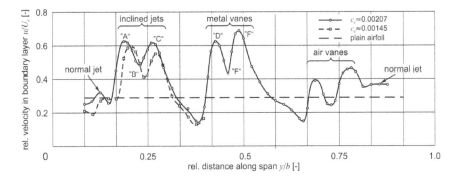

**FIGURE 6.14**  Spanwise velocity distribution in boundary layer at fixed distance from surface $x/c = 0.875$, $\alpha = 8°$ affected by fluidic vortex generator jets; $U_s$ – a velocity in the boundary layer as measured by a surface total head probe. (According to Wallis [28] figure 3.)

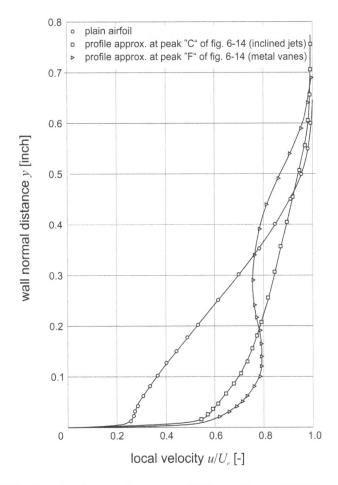

**FIGURE 6.15**  Boundary layer profiles at $x/c = 0.875$ ($\alpha = 8°$. $c_q = 0.00207$) controlled by fluidic vortex generator jets. (According to Wallis [28], figure 7.)

### 6.1.5 FLUIDIC OSCILLATOR

Independent of the flow control mechanism, a time-varying actuation shows benefits over steady actuation, at least in terms of the required mass flow. For wind tunnel and laboratory tests, either fast switching valves (e.g. [19]) or loud speakers (e.g. [23]) have been applied to attenuate the flow velocity of the jets. Such systems are not suited for an aircraft due to their weight and/or limited long-term reliability. For this reason, the research on suitable actuators that provide a pulsed airflow is of a similar importance as the understanding of the actuated flow to enable AFC on aircraft.

An important invention was made by Warren [31] with the fluidic oscillator shown in Figure 6.16. The main flow of the oscillator passes from the pressure supply (no. 15)

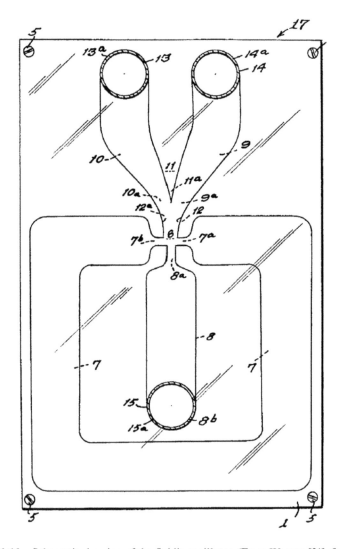

**FIGURE 6.16**  Schematic drawing of the fluidic oscillator. (From Warren [31], figure 1.)

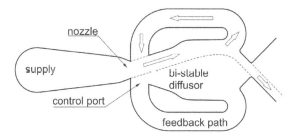

**FIGURE 6.17** Mechanism of oscillation in bi-stable fluidic oscillator. (According to Graff et al. [32], figure 7 Left.)

to a bi-stable nozzle (9/10). The bi-stable state is obtained by a diffuser that is too steep and does not allow for attached flow to the side walls. Therefore, the air flow stable attaches to one side wall of the diffuser, while the flow on the opposite side is separated. As the flow channel is symmetric, there is no condition for the air flow to attach at a specific side. The oscillator is created by introducing the feedback lines (no. 7). Figure 6.17 shows the flow pattern inside the actuator. When the flow attaches to the wall, the feedback line gets a pressure change that travels with the speed of sound through the feedback line. At the end, the disturbance in the feedback line changes the local pressure at the main air flow, causing the flow to attach to the opposite side. The frequency of the process is hereby depending on the flow properties (pressure, density of air supply), the mass flow rate, and the geometric length of the feedback line.

The bi-stability of the actuator causes the switching to happen very suddenly. At the end, the exiting jet "switches" from one side to the other. If both channels are distinctly separated by a diverting wall, as at Warren's fluidic oscillator, the actuator provides two pulsed jets with a phase shift of 180 degrees. In case of no diverting element, the device is a so-called sweeping jet [32], as the jet "sweeps" from left to right but not being intermittent by itself. Anyhow, also the sweeping happens very suddenly and there is no difference in the principal aerodynamic effect. Only the shaping of the nozzles may provide more easily a defined jet direction in case of the fluidic oscillator, while the sweeping jet has the tendency to spread out slightly in the direction of the span of the nozzle.

### 6.1.6 Synthetic Jets

As an alternative to provide pressurized air for flow control from a remote source, like, e.g., the aircraft engine, synthetic jet actuators generate the needed airflow directly at the location where flow control is applied. As there is no externally provided mass flow, they are also referred to as "Zero Net Mass Flux" (ZNMF) actuators. The driving energy source is mostly electric, in few cases hydraulic.

Initial attempts to provide a pulsating air flow used loud speakers emitting a harmonic pressure variance into a reservoir, which was connected by a simple orifice or slot to the flow [33, 34]. Such concepts were appropriate for investigating the principal effects in wind tunnels but are no system to be placed on an aircraft.

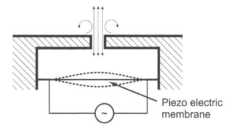

**FIGURE 6.18** Principle of a synthetic jet actuator based on employing a Piezo-electric ceramics membrane.

Amitay et al. [35] introduced more promising devices using oscillating Piezo-electric ceramics membranes, as sketched in Figure 6.18. The oscillating membrane compresses and expands the cavity volume. The small orifice injects flow into the outer flow to be controlled during the high-pressure phase. When the generated air-jet leaves the orifice, a secondary ring vortex structure can be observed in still air. As a further consequence, the air flow to be provided as blowing has to be sucked into the actuator before compression. The advantage of Piezo ceramics over solenoids-driven loud speakers is their capacitive resistance. To drive the actuator, there is a significant voltage needed, but currents are low.

A current issue in actuator development is the capability of the synthetic jet actuators to provide air flow with sufficiently high exit velocity. Gallas et al. [36] reported on the key design parameters to design a highly efficient synthetic jet actuator. The actuators are characterized by two characteristic frequencies. One is the Helmholtz resonance frequency of the cavity volume, the second is the eigenfrequency of the membrane. It has been shown to be important that these two frequencies match to achieve an overall resonance leading to a high exit velocity. Synthetic jet actuators reported by Liddle and Wood [37] achieved an exit velocity of about 50 m/s. Later, Gomes et al. [38] reported on synthetic jet actuators that achieved a peak exit velocity of up to 130 m/s. Nevertheless, the current potential of synthetic jet actuators is still less than those of externally pressurized air-driven pulsed jet actuators, where the jet velocity is a direct function of the provided pressure and can easily achieve sonic speeds.

## 6.2 CIRCULATION CONTROL

The relation between lift and circulation given by Zhukovsky's theorem has been explained in Section 2.1. Combining eqs. (2.21) and (2.22) leads to a definition of the sectional lift coefficient based on the circulation

$$c_L = \frac{2\Gamma}{U_\infty c}. \qquad (6.21)$$

Additionally, it was discussed that, in attached incompressible inviscid flow, the amount of circulation is defined by the Kutta-condition to fix the rear stagnation point at the trailing edge. Anyhow, this discussion did not include the possibility to alter the velocity at the trailing edge by injecting flow at a higher total pressure than the free stream. To keep the rear stagnation point at the trailing edge, the circulation has to change in order to compensate for this disturbance. As in eq. (6.21) the reference velocity is still related to free stream condition, the change in circulation is directly attributed to the change of sectional lift coefficient. Lan and Campbell [39] define the additional lift due to blowing

$$\Delta L_j = \dot{m} \cdot u_j \sin(\alpha + \delta_j)$$ (6.22)

accounting for the component of the thrust force in the direction of lift. Expressed in terms of additional circulation [40] it reads[2]

$$\Delta \Gamma_j = \frac{\dot{m} u_j}{\rho_\infty U_\infty b}(\alpha + \delta_j)$$ (6.23)

or expressed with the momentum coefficient[3]

$$\Delta \Gamma_j = \frac{1}{2} c_\mu \cdot U_\infty \cdot c(\alpha + \delta_j)$$ (6.24)

as denoted by Spence [42]. Further analysis of the potential flow takes into account the downwash induced by a jet emanating at a certain inclination angle at the trailing edge of an airfoil. In general, the lift increase is dependent on the actual angle of attack and jet inclination

$$\Delta C_L = \frac{\partial C_L}{\partial \delta_j} \delta_j + \frac{\partial C_L}{\partial \alpha} \alpha.$$ (6.25)

According to Maskell and Spence [43], a closed relation of the two derivatives is given by

$$\left(\frac{\partial C_L}{\partial \delta_j}\right)^2 = 2c_\mu \left(\frac{\partial C_L}{\partial \alpha}\right) - c_\mu^2$$ (6.26)

and approximation formulae are given by

$$\frac{\partial C_L}{\partial \delta_j} = 2\sqrt{\pi c_\mu}\sqrt{1 + 0.151\sqrt{c_\mu} + 0.139 c_\mu}$$

$$\frac{\partial C_L}{\partial \alpha} = 2\pi\left(1 + 0.151\sqrt{c_\mu} + 0.219 c_\mu\right).$$ (6.27)

## 6.2.1 THE COANDĂ EFFECT

"It is an observed fact that when a stream or sheet of fluid issues through a suitable orifice, into another fluid, it will carry along with it a portion of the surrounding fluid, if its velocity is sufficient. In particular, if a sheet of gas at high velocities issues

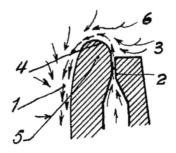

**FIGURE 6.19**  Schematic drawing of the Coandă-Effect: the fluid ejected through a nozzle (2) entrains the outer fluid (3) and stays attached to the outer surface of a body (5) along the curved surface. (From Coandă [44], figure 3.)

into an atmosphere of another gas of any kind, this will produce, at the point of discharge of the said sheet of gas, a suction effect, thus drawing forward the adjacent gas. If at the outlet of the fluid stream or sheet, there is set up an unbalancing effect on the flow of the surrounding fluid induced by said stream, the latter will move towards the side on which the flow of the surrounding fluid has been made more difficult" – (Coandă [44] p. 1).

With these words, Henri Coandă describes the effect that was named after him within the US patent [44] he obtained one year after describing the effect in his French patent one year before. Later, he describes the effect as a "stream of gas which passes through [a] slot has the tendency to follow the surface [...]. In this way it is possible [...] to direct a large volume of ambient fluid by a small mass of fluid under pressure" – (Coandă [45] p. 8). Figure 6.19 of his original patent shows the principal effect. Explanations that can be found in encyclopedias define the Coandă effect as "the tendency of a jet of fluid emerging from an orifice to follow an adjacent flat or curved surface and to entrain fluid from the surroundings so that a region of lower pressure develops" – (Merriam Webster Dictionary [46]).

It is important to recognize the developing low pressure by entrainment to be responsible for the attachment of the jet flow. So, the Coandă effect is always bound to the presence of a jet of fluid at a higher total enthalpy than the surrounding flow. The suction pressure developing on the surface is the necessary force counteracting the centrifugal force of the jet.

### 6.2.2  CIRCULATION CONTROL AIRFOIL

The concept of an airfoil completely relying on the Coandă effect for lift generation was promoted by Englar [47]. Figure 6.20 depicts the airfoil and sketches of the flow field without blowing and with blowing at low and high pressure. The airfoil is characterized by a blunt rounded trailing edge with a blowing slot at the upper side. Pressurized air is emitted as a jet through the slot and follows the circular surface due to the Coandă effect. As the jet travels around the trailing edge, it is suspect to friction forces at the surface and at the shear layer to the outer flow. By this, the jet is decelerated and the suction pressure counteracting the centrifugal forces reduces. With a medium supply pressure at some point along the surface, the energy of the

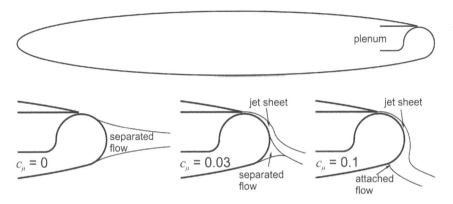

**FIGURE 6.20**  Blunt rounded trailing edge with blowing slot on upper side to fully control the generation of lift by the Coanda effect. (Up: According to Englar [47], figure 12; Low: Sketched after Radespiel et al. [48], figure 10.)

jet is consumed so there is no additional suction force, and the jet sheet leaves the surface. The location of the flow separation is mainly related to the supply pressure of the air jet in relation to the static pressure of the outer flow field.

At a certain pressure level, the jet is not detaching before it hits the lower surface flow forming a common stagnation point. In this condition, the trailing edge shows no longer a separated boundary layer flow. Increasing the pressure level further moves the common rear stagnation point upstream along the lower airfoil side. This direct change of the rear stagnation point alters the Kutta condition further, so directly influencing the circulation generated by the airfoil.

There is a question arising in comparison to the tangential blowing discussed in Section 6.1.2: What's the difference? The difference is the applied pressure level and the effectivity. Englar [49] gives a good explanation by looking at the lift increment in dependence of the momentum coefficient of a Coandă airfoil.

Figure 6.21 resumes the lift increment variation due to the introduced jet momentum. The first regime of tangential blowing through the slot is associated with a high effectivity. This is related to reducing the separated flow in the trailing edge region in terms of **boundary layer control** by the mechanism described in Section 6.1.2.

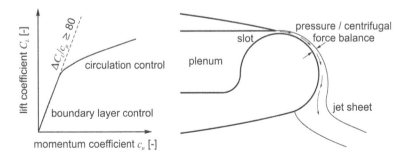

**FIGURE 6.21**  Efficiency of lift generation of Coandă blowing. (According to Englar [49], figure 1.)

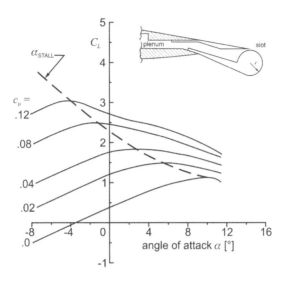

**FIGURE 6.22** Lift curves at constant momentum coefficient for a NACA 66-210/CCW airfoil. (Reproduced from Englar [50], figures 4 & 14(a).)

In this regime, the momentum of the jet flow is capable to compensate for the viscous losses of the boundary layer. Once the separation is completely suppressed, the effectivity drops. There is still an increase of lift coefficient but now induced by **circulation control** using the Coandă effect. This regime is characterized by a jet momentum that is significantly higher than the incoming flow according to high momentum coefficient values. It is important to note that the general relation of the lift coefficient and the blowing momentum coefficient shown in Figure 6.21 is found in the same or similar way for all kinds of tangential blowing applications. In case of non-circular trailing edges, the circulation increase is then defined by the angle under which the jet is leaving the airfoil and the relations given in eq. (6.27).

One common observation of circulation control airfoils is the enormous increase of lift coefficient. Englar reports lift coefficients close to $C_{L,\max} = 7$ for a 17% thick transonic airfoil with a small Coandă trailing edge [49]. But a drawback of producing the lift at the trailing edge by circulation control is the corresponding movement of the stagnation line at the airfoil leading edge, producing a high suction peak at already low angles of attack. Figure 6.22 shows the change of the lift characteristics depending on the momentum coefficient from investigations with a NACA 66-210 airfoil with circulation control [50]. A decrease in the stall angle of attack into the negative regime is obvious. In order to make use of circulation control at the trailing edge, an additional leading edge high-lift system is necessary to shift the stall incidence back into the positive regime.

### 6.2.3 COANDĂ FLAP

A distinct shortcoming of the Coandă airfoil is the constant need of blowing to provide a significant level of lift coefficient. Without the blowing, there is no distinct

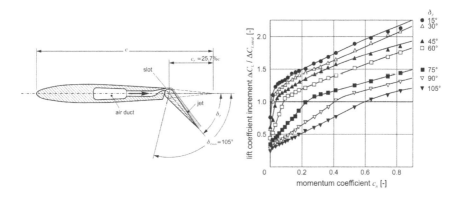

**FIGURE 6.23** Coandă flap or plain blown flap and the corresponding lift increment related to theoretically ideal flap efficiency. (According to Thomas [52], figures 2 & 7(b).)

trailing edge Kutta condition and the lift generated is arbitrarily low. Building an aircraft relying completely on such a system, with the according impact on flight behavior in case of failure, is critical and needs a high system effort to guarantee reliability. It is, therefore, worth to look for configurations that allow for the usage of the Coandă effect, which does not need blowing at low but significant lift coefficients, e.g., in cruise flight, and with less critical impact on the flight handling in case of failure.

Such a concept is the Coandă flap, as illustrated in Figure 6.23 (left). In cruise flight, the Coandă flap is retracted and a normal airfoil is formed. The Coandă flap is designed for very high deflection angles in the range of $\delta_F = 65° \div 80°$. When deflected, a slot is opened and a jet flow is provided from a pressurized plenum inside the wing. Without blowing, the flap flow would be separated as discussed in Section 5.1.1. The Coandă flap is similar to a boundary layer control tangentially blown plain flap as discussed in Section 6.1.2 but at a higher blowing rate. The provision of the jet flow uses the Coandă effect to keep the flap flow attached [51].

As for the pure Coandă airfoil, also here we obviously have to distinguish between boundary layer control and circulation control, visualized in Figure 6.23 (right). It shows the lift increment due to blowing related to the theoretical lift increment due to flap deflection according to eqs. (5.1) & (5.3). Boundary layer control is achieved in the range of blowing momentum where the flap separation is reduced in its size. Once the flap flow is completely attached, the transition is towards circulation control. This can be seen in the figure when the gradient of the lift increment reduces. It is not astonishing that this is close to the theoretical lift increment as the blowing compensates for all viscous losses. After achieving the attached flap flow, the additional momentum is active in the sense of thrust vectoring according to the discussion at the beginning of this section. (see eq. (6.27)) with the jet inclination angle, $\delta_j$, being the angle of the deflected flap upper surface at the trailing edge.)

There have only been built a few aircraft implementing a Coandă flap, most of them in the military sector. The Huntington H.126 was a technology demonstrator and achieved a stall speed of 32 mph (51 km/h). The first production aircraft to implement a Coandă flap was the Lockheed F-104 Starfighter. Other military aircraft are the F-4 Phantom and the Buccaneer. Another impressive example of an aircraft is the British Aircraft Cooperation BAC TSR2, although this aircraft never went into serial production. With the Coandă flap, the take-off field length was reduced from 6000 ft (1800 m) to 1600 ft (490 m). Nevertheless, the most impressive Coandă flap aircraft is the Shin Meiwa used in a civil variant for Search & Rescue missions [53]. It utilizes a separate 1 MW gas turbine inside the fuselage to provide the pressurized air for the Coandă flap.

### 6.2.4 FLUIDIC GURNEYS

Another concept of circulation control is the replacement of a rigid Gurney flap (see Section 5.1.3) at the trailing edge by a fluidic device generating a similar effect on the airfoil flow [54]. The fluidic Gurney is achieved by blowing out pressurized air orthogonally to the surface at the lower side of the airfoil near the trailing edge. Figure 6.24 shows the flow field and the pressure distribution of an OAT15A airfoil

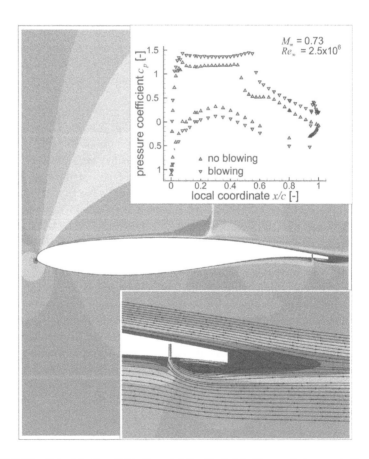

**FIGURE 6.24**   Principle flow field of an airfoil with fluidic Gurney in transonic flow.

with Fluidic Gurney in transonic flow. Similar to the Gurney flap, the blowing forms a disturbance near the trailing edge that relieves the Kutta condition. The effect can be seen in the pressure distribution as difference in the pressure level on the upper and lower sides of the airfoil near the trailing edge.

Current research with this device focuses on its application for buffet prevention in transonic flow. Nevertheless, this type of influence on the outer flow has proven to effectively provide the same effect as a rigid Gurney, although at quite high blowing rates in the order of $c_\mu \approx 1 \div 5\%$. The advantage is that it can be switched off whenever it is not needed and, therefore, getting rid of a certain drag penalty. The drawback is the challenging structural integration into the slim wing trailing edge.

## 6.3  POWERED LIFT

Thrust supported high-lift covers the direct usage of the high momentum of the engine jet to generate a significant increase in lift. Sometimes, also the usage of high-pressure air from the engine compressor stage for AFC is summarized under this topic. Also, the direct generation of lift force by redirecting the engine thrust – so-called thrust vectoring – may be included. Anyhow, these two topics are nevertheless excluded in the discussion, as the first has been discussed already, and the second is at the border to vertical take-off aircraft that are not in the scope of this book. Here, we concentrate under this topic on the usage of the engine exhaust jet to provide a higher wing circulation.

### 6.3.1  PROPULSIVE SLIPSTREAM DEFLECTION

One of the early developments on Short Take-Off and Landing (STOL) capability has been conducted in France. The Breguet Aircraft Company started in 1945 the design work on a military transport aircraft for short field operations. The Breguet 940, first flight in 1957, incorporated some unique features that made it a good basis for a STOL aircraft prototype, the Breguet 941. This aircraft was flight tested and analyzed by NASA [55] in the early 1960s "in order to gain additional information on the operation of large STOL aircraft" – (Quigley et al. [55] p. 2). The aircraft was equipped with four turboprop engines with the propeller slipstream almost extending over the full wing span, as seen in the plan view in Figure 6.25 (left). The propellers on each side were counter-rotating and were equipped with an inter-connect, avoiding the negative impact of an engine failure. The triple-slotted flaps, also spanning the full wing span, were deflected by 98° inboards (Figure 6.25 (right upper)) and 65° outboards (Figure 6.25 (right lower)).

Figure 6.26 reproduces the operational envelope of the Breguet 941 aircraft. The usual approach speed of approximately 60 kts flown in the flight tests resulted in a glide path angle of $5° \div 7.5°$ at about half power. Anyhow, the operational envelope shows that a go-around in landing configuration was only possible at full thrust. As a safe operation for one engine inoperative (OEI), therefore the flap setting had to be reduced to 75° inboards and 50° outboards. "The take-off and landing distances over 35 and 50 feet obstacles, respectively, were less than 1000 feet" – (Quigley et al. [55] p. 1).

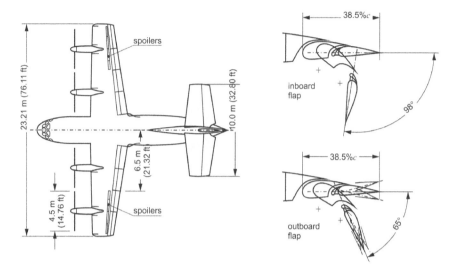

**FIGURE 6.25** Plan view of the Breguet 941 STOL aircraft (left) with 38.5% triple-slotted flaps deflected inboards at 98° (middle) and outboards at 65° (right) – (According to Quigley et al. [55], figures 2 & 3.)

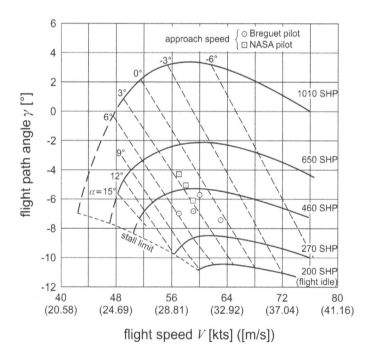

**FIGURE 6.26** Operational envelope of the Breguet 941 STOL aircraft. (According to Quigley et al. [55], figure 10.)

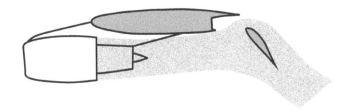

**FIGURE 6.27** Schematics of the flow field of an externally blown flap. (According to Johnson [58], figure 1.)

Propeller slipstream deflection recently got attention again for designs of efficient short-range aircraft capable to land on short airfields nearby urban settlements. Such aircraft would require STOL capabilities as well as low noise footprints. Beck et al. analyzed numerically and experimentally a wing-propeller combination with an additional Coandă flap [56]. Burnazzi and Radespiel added a smart droop nose device (see Section 5.1.4) to the leading edge in order to compensate for the low stall incidence with a gap-less high-lift system [57].

### 6.3.2 EXTERNALLY BLOWN FLAPS

On larger civil transport aircraft with more than 100 passengers, the engine installation below the wings has turned out to be favorable against rear fuselage mounted engines. These are nowadays seen for smaller aircraft only (70 passengers and less). The mounting of the engine in front and below the wing causes an interaction of the engine jet with the flow on the pressure side of the wing. If the engine is coupled close enough to the wing, the engine jet may impact the flap system. Figure 6.27 sketches the principal arrangement of high-lift wing and engine. Using the higher momentum of the jet for an increased generation of lift from the flap due to the higher dynamic pressure is the so-called externally blown flap.

Figure 6.28 illustrates the two-fold effect of the externally blown flap by the lift increment depending on the engine thrust. First, there is the direct effect of turning the thrust vector into lift direction[4]. The contribution of the thrust vector depends on the angle of attack, as the thrust axis is usually close to the aircraft axis. The thrust vector is deflected by the flap and therefore results in a jet inclination. Second, the flap is exposed to a higher flow momentum, thus creating more lift force at the same incidence symbolled in Figure 6.28 by $C_{L,\Gamma}$.

The turning of the jet depends on the penetration depth of the flap into the jet. "Direct application of the jet flap theory to the externally blown flap would assume that the jet turning angle, $[\delta_j]$, is equal to the flap deflection angle, $\delta_F$ This has proved to be a good assumption for the case where the flap system captures the entire jet efflux; however, where the flap intercepts only a part of the jet, the effective turning angle will be less than the flap angle" – (Roe [59] p. 6). Roe defines an impingement parameter based on geometrical relations as shown in Figure 6.29. The jet diameter at the end of the flap is approximated by

$$\frac{R}{D_j/2} = 1 + \frac{x/D_j}{2.3}.$$

(6.28)

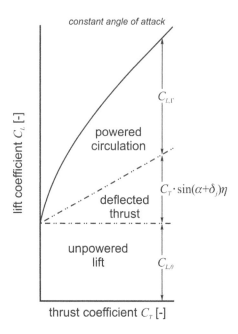

**FIGURE 6.28** Contributions to the lift increment induced by externally blown flaps. (According to Johnson [58], figure 3.)

From measurements, the dependency of the jet turning on the impingement parameter has been derived as shown in Figure 6.30. "Test data show nearly complete jet turning with flap immersion of $z_F/R > 0.65$ except for cases where flow separation over the flaps is suspected" – (Roe [59] p. 8). With this ratio, the thrust direction $\delta_j$ can be estimated from the flap deflection angle $\delta_F$ and used in eq. (6.25) to calculate the impact on lift and drag coefficients.

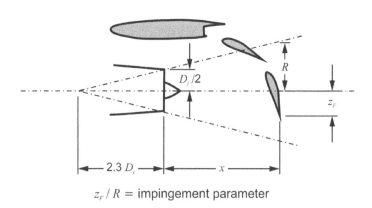

$z_F / R$ = impingement parameter

**FIGURE 6.29** Definition of flap impingement parameter according to Roe. (According to Roe [59], figure 2.)

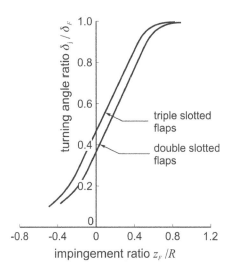

**FIGURE 6.30**  Jet turning angle depending on the impingement parameter. (According to Roe [59], figure 6.)

Another approach to quantify the ability of a blown flap to turn the thrust vector is achieved by the thrust recovery factor. Johnson [58] established it as the ratio of the resulting total force to the engine thrust

$$\eta_j = \frac{\sqrt{F_{normal}^2 + F_{axial}^2}}{T}. \tag{6.29}$$

To quantify the thrust recovery ratio, measurements with engine simulators blowing on flaps were performed in quiet air. It defines the efficiency of the flap for reorienting the thrust vector. In ideal case, the flap fully deflects the engine thrust without losses. Figure 6.31 shows a summarizing diagram of the thrust recovery factor[5] obtained from measurements with different engine types and flap systems. It shows that, even for very high flap deflections, a thrust recovery factor of up to 80% is achievable.

Figure 6.32 shows measurement data of wind tunnel test on a blown flap configuration [58]. It displays thrust-corrected lift and drag values, where the direct contribution of the thrust by thrust vectoring is subtracted from aerodynamic coefficients

$$C_{L,T-corr} = C_L - C_T \sin(\alpha + \beta_j)\eta_j$$
$$C_{D,T-corr} = C_D + C_T \sin(\alpha + \beta_j)\eta_j \tag{6.30}$$

and the dashed line shows the theoretical polar obtained from basic assumptions based on the ideal induced drag only

$$C_D = \frac{C_L^2}{\pi\Lambda}. \tag{6.31}$$

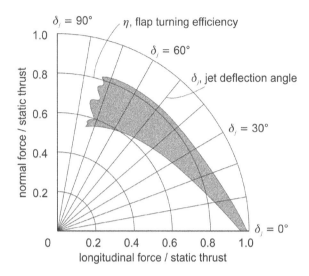

**FIGURE 6.31**   Thrust recovery factor depending on jet deflection. (According to Johnson [60], figure 4.)

"The data show that these collapsed power-on thrust-removed lift-drag polars follow the same trend as the traditional induced-drag parameter" – (Johnson [58] p. 45).

The most prominent example of an EBF aircraft was the McDonnel Douglas YC-15 [61]. It has been built in the scope of the US Air Force competition on the Advance Medium STOL aircraft program initiated in 1971. Although the program was stopped after the prototype demonstrations, later on, McDonnel Douglas used the EBF concept for the development of the C-17 Globemaster [62].

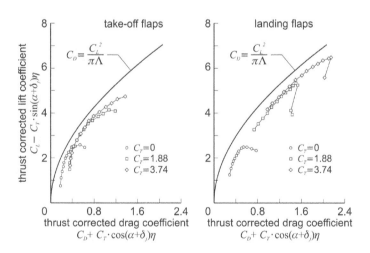

**FIGURE 6.32**   Effect of increasing thrust coefficient on the thrust-removed lift-drag polars. (According to Johnson [58], figure 6.)

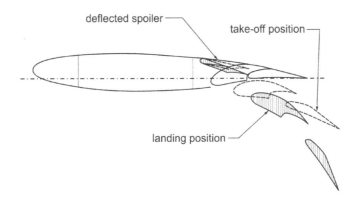

**FIGURE 6.33**  Flap system of the McDonnel Douglas YC-15. (According to Heald [61], figure 3.)

The design requirements were a turbojet powered aircraft for a 400 NM range and an operational field length requirement of 2000 ft (609 m). The YC-15 was equipped with four Pratt & Whitney JT8D-117 engines mounted under the wing. The selection of a four engine configuration was mainly driven by the one engine out considerations. To improve the efficiency of the EBF, the flap system consisted of a double slotted flap that was deflected deeply into the engine jet path (Figure 6.33). To achieve the correct gap flow, the spoilers were drooped to achieve closer gaps at lower flap positions.

Figure 6.34 reproduces the operational envelope of the McDonnel Douglas YC-15 in approach condition. The normal approach condition with 65% engine thrust

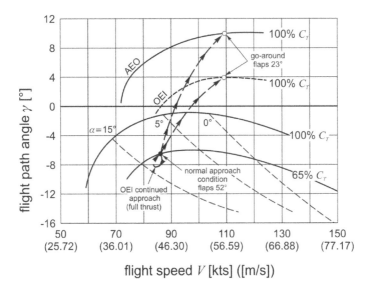

**FIGURE 6.34**  Operational envelope of the McDonnel Douglas YC-15 in approach condition. (According to Heald [61], figure 11.)

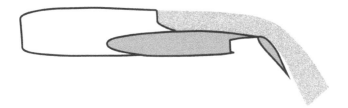

**FIGURE 6.35**  Principal arrangement of an upper surface blowing aircraft.

achieved a static glide path of 6.6° at 65 kts approach speed. It is seen that the aircraft
was not able to climb even at full thrust and for a go-around the flap setting had to be
decreased from 52° to 23°. The approach thrust had been limited to 65% to "provide
a margin for go-around after loss of an engine" – (Heald [61] p. 4). The recovery path
for go-around with all engines or engine failure is also seen in the figure.

### 6.3.3  Upper Surface Blowing (USB)

An alternative method to directly use the momentum of the engine jet for lift generation is to place the engine above the wing and to tangentially blow along the wing
upper surface. As illustrated in Figure 6.35, the engine jet follows the wing contour
by the Coandă effect. In order to magnify the lift generation effect, it is beneficial
to use a Coandă surface at the trailing edge instead of a slotted flap, as the gap flow
would reduce the bending of the engine jet into the lift direction.

"A number of critical parameters are involved in effective jet turning; among
them is the ratio of jet thickness to turning radius" – (Phelbs [65] p. 99). Detailed
analysis has been made by Sleeman and Phelbs [66]. It was shown that a flatter jet
is better able to follow the wing surface than the exit of a circular nozzle. Yen [63]
reports the flow turning angle for nozzles of different aspect ratios depending on the
ratio of the flap radius to the nozzle height. Figure 6.36 shows the summary of those

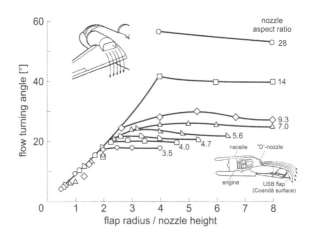

**FIGURE 6.36**  Effect of engine nozzle size and aspect ratio on flow turning angle. (According
to Yen [63] figure 6 and Riddle & Eppel [64] figures 3 & 7.)

measurements. It reflects that in general, the higher nozzle aspect ratio provides better flow turning. For small flap radii to nozzle heights, the jet is sufficiently thin and a
higher nozzle aspect ratio will improve the lift generation due to the larger impacted
area, but not due to a better flow turning.

The effectiveness of upper surface blowing has been demonstrated in the 1970s.
In the scope of the already mentioned US Air Force competition on the Advance
Medium STOL aircraft (AMST) program, Boeing proposed and developed a USB
configuration, the Boeing YC-14 [67]. Two engines are mounted ahead and above the
wing upper surface close to the fuselage to mitigate high yawing moments in case of
engine failure. The inboard flap was a Coandă surface deflected at 50°, while the outboard flap was selected as a double-slotted flap. The leading edge was equipped with
Krueger flaps and an additional BLC system at the leading edge over the full span.
"The performance with all engine operations was especially outstanding. It could
land at 65 knots in about 1150 ft and take-off in a ground roll of 1,100 feet" – (Nark
[67] p. 30-3). Figure 6.37 resembles the operational envelope of the Boeing YC-14 in
approach configuration. The approach at the targeted glide path of 6° is indicated in
the graph. Like for other powered lift aircraft, this approach condition was slightly
at the "back side" of the polar, implying that a commanded thrust increase results in
a speed reduction due to increasing the lift force. In order to "provide the feeling on
the front side the polar, – the way he was fundamentally trained to fly. [...] This was
accomplished by developing a computer controlled, triply redundant, electrical flight
control system (EFCS). [...] The pilot had a very simple task to master on approach.
Use the stick to point the nose of the airplane where he wanted to land. Everything
else was taken care of for him by the EFCS" – (Nark [67] p. 30-5).

At NASA, the Quiet Short-Haul Research Aircraft (QSRA) program was started
in 1976, in parallel to the AMST program [68]. It led to the development of a four

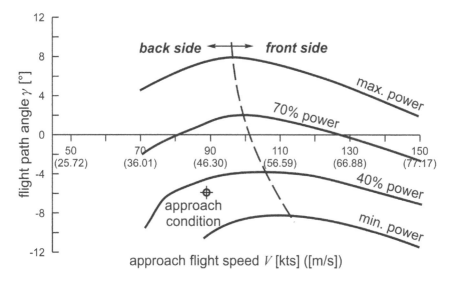

**FIGURE 6.37**   Operational envelope of the Boeing YC-14. (According to Nark [67],
figure 16.)

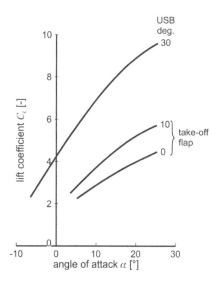

**FIGURE 6.38** Lift coefficients obtained for the upper surface blowing aircraft NASA QSRA. (According to Queen & Cochrane [70], figure 3.)

engine upper surface blowing configuration based on a deHavilland C-8A Buffalo. "This configuration permits a larger span USB flap and reduces adverse yaw and roll moments with one engine inoperative, thus yielding the improvement in lift over the twin-engine YC-14" – (Shovlin & Chochrane [68] p. 15).

Figure 6.38 shows lift coefficients obtained by NASA during flight tests with the Quiet Short-Haul Research Aircraft, which are not thrust-corrected and include the contribution of thrust vectoring. Nevertheless, a value of $C_L \approx 10$ for the fully deflected Coandă surface shows the efficiency of the jet turning. "FAR take-off field lengths can be shortened approximately 30% by use of powered lift USB technology as compared to conventional aircraft with the same thrust-to-weight ratios and wing loadings" – (Riddle et al. [69] p. 12).

In the late 1970s, the National Aerospace Laboratory on Japan started a program for a quiet STOL aircraft development [71] based on a Kawasaki C-1 aircraft. The project started in 1977 and the aircraft was successfully flight tested between 1985 and 1989, with a total of 97 flights and 170 flight hours. To achieve the STOL requirements, the aircraft was designed as an Upper Surface Blowing configuration with four newly designed turbofan engines and additional Boundary Layer Control at the wing leading edge and the ailerons. The inboard flap designed as Coanda surface for the engine exhaust was deflected by 60°, the outboard flap by 65°. In-flight tests, the aircraft demonstrated a landing field length of 1473 ft (449 m) and a take-off distance of 1670 ft (509 m), both counted for clearing a 35 ft obstacle.

Figure 6.39 reproduces the reported flight path angle versus flight speed for the landing configuration of the ASUKA QSTOL research aircraft. The authors comment that "the powered lift STOL aircraft has naturally the decoupling V-Gamma characteristics, that is, with a constant pitch attitude flight path angles can be control only by the engine power" – (Mori et al. [71] p. 1273). Another specific difficulty of

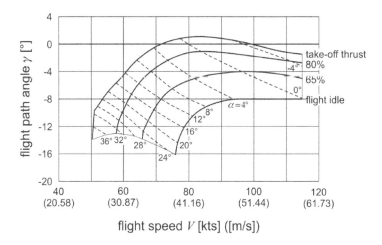

**FIGURE 6.39** Operational envelope of the NAL ASUKA QSTOL research aircraft in landing configuration. (According to Mori et al. [71], figure 18.)

the configuration is with regard to ground effect. It is reported that the ground effect is effective at altitudes less than $h/b < 0.4$. "[...] the lift is increased about 10% of the lift in free air, the drag is decreased, the downwash angle is also decreased. [...] This decrement of the downwash angle generates large nose down moment change at the touch down. The positive lift increments make it difficult to land at a pin point" – (Mori et al. [71] p. 1273).

The only upper USB configuration that was produced in series is the Ukrainian Antonov AN-72. The configuration looks similar to the Boeing YC-14. The differences are mainly a swept wing and a different leading edge, as the AN-72 is equipped with slats rather than Krueger flaps and doesn't implement additional BLC at the leading edge. The first flight was conducted in 1977 [72]. Totally, 114 aircraft were entered into service from 1984 to 1992. "The AN-72 is a light short takeoff and landing jet transport airplane intended for operations on non-equipped air strips 600 m long" – (Antonov Company [73]).

## 6.4 A REMARK ON THE APPLICATION OF ACTIVE HIGH-LIFT FOR CIVIL TRANSPORT AIRCRAFT

AFC technologies are not yet implemented on civil transport aircraft on a regular basis. Only one civil transport aircraft (Antonov AN-72), one military transport aircraft (McDonnel Douglas C-17), and one amphibium (ShinMeiwa US-1) ever went into serial production. The aerodynamic effectivity has been proven, but the technology has not yet found its way into application. Besides several other reasons to withdraw from upcoming technologies that may not yet have proven their robustness, there are few factors majorly responsible for a significant reluctance of aircraft designers to fully rely on AFC systems or powered lifts. There are a four questions that have to be answered before active high-lift technology will be able to show its full benefits in civil commercial air transport.

### Is the Effort Worth the Benefit?

Many studies have seen AFC as a benefit of its own neglecting the effort to be taken to provide an efficient system. Most studies only report on the aerodynamic benefit, some of them at least mentioning the needed effort of introduced momentum. More applied studies tried to incorporate the effort in an approximate form by taking into account the energy needed, e.g., to provide pressurized air. One of the meaningful characteristic values is the aerodynamic figure of merit defined by Seifert [74]

$$AFM_1 = \frac{(L/D)}{\left(L/\left(D + P/U_\infty\right)\right)}.\tag{6.32}$$

Nevertheless, the additional effort for supporting systems with their impact on weight, complexity, and maintenance is often not addressed. Still "a common precept within the aeronautical community is that powered-lift technology is only applicable to short takeoff and landing (STOL) aircraft and that such aircraft have high thrust-to-weight ratios and, therefore, are inefficient" – (Riddle et al. [75] p. 1). Anyhow, the realization of powered lift configurations and especially "[...] flight research with the Quiet Short-Haul Research Aircraft (QSRA) has confirmed that landing performance at relatively short field lengths can be achieved at the lower thrust-to-weight ratios comparable to those used in conventional aircraft" – (Riddle et al. [75] p. 1). At the end, the full benefit of AFC or powered lift can only be exploited if the aircraft design incorporates those technologies from the very beginning. Establishing the evaluation of these technologies on a retro-fit basis as it is mostly done is not able to unveil the technological potential.

### Is the System Safe and Reliable?

An AFC system must be highly safe and reliable in order to rely on the benefit in the most critical airfoil state, somewhere close to stall. This implies a systematic redundancy design, as the aircraft has to rely on the presence and functionality of the system to be airworthy. In one-engine-out-condition, this is even more critical, as to achieve the required climb gradient probably the full engine power is needed at the same time for thrust and for guaranteeing a safe flight state. As seen for powered lift configurations, where approximately 50% of propulsive force is used for lift generation, the failure of an engine can get critical, mostly avoided by reconfiguring the aircraft. Anyhow, past and recent studies, e.g., by Keller [76], show that the unbalance of engine failure can be contained if addressed from the beginning by incorporating sophisticated engine controls and cross feeds to avoid the one-sided flow control failure at the same time.

### Is the Energy Need and Availability Balanced?

When applying an AFC system, it is necessary to assess the availability of the energy source in the case the availability of the AFC system is critical for flight. This holds both for pressurized air, which most likely is taken from the engine bleed air, and the electric power to be produced in addition by the generator, mostly also driven by the same engine. In take-off, the power off-take reduces the available thrust for climbing, and the benefit of the flow control has to be properly accounted for. In landing, usually, engines should be on idle mode. If there is power to be used, it has to be guaranteed that increasing the power setting does not result in excess thrust that spoils the glide slope of the aircraft.

A specific challenge of powered lift configurations is the landing on the "back-side" of the power curve. On "back side," increasing thrust results in increasing lift and thus reduced flow speed – the opposite of the aircraft behavior expected by pilots. Loth [77] noted that "flying slow, on the backside of the power curve, is not recommended because no power is then left over to assist in stall recovery. [...] Until one discovers a new technology which prevents having to land on the backside of the power curve, circulation control may not become popular with pilots for take-off and landings" – (Loth [77] p. 613). The Boeing YC-14 demonstrated that this can be handled with an elaborated EFCS system.

### Will the System Ever be Used?

Referring back to Section 3.2.3, the operational limits of flight are offset from the physical limits by the safety margins on minimum and maximum speeds. Figure 6.40 revisits the visualization of the operating limits by the

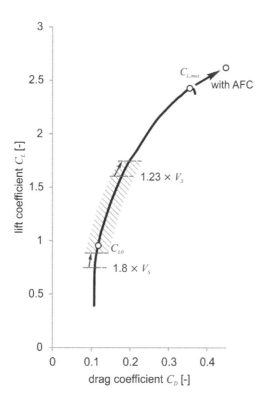

**FIGURE 6.40**  Aerodynamic force polar showing the shift of the operational range induced by the usage of stall preventing active flow control systems.

aerodynamic force coefficients shown in Figure 3.3 by the shift of the maximum lift coefficient obtained by AFC. It is seen that the operational limits follow the maximum lift increment. On the other side, the limits are still in a range of aerodynamic properties, where the aircraft is able to fly from a physical perspective. The AFC system is therefore not needed to produce the lift but to reinstall the safety margin against stall. This means that it is, to some extent, doubtful that the AFC system – although necessary to be **allowed** to fly – needs to be switched on to be **able** to fly. In the most extreme case, it is imaginable to have an AFC system installed on an aircraft to recapture safety margins when needed, which is complex to maintain and brings some weight penalties, and may never be switched on in normal flight procedures – and probably not during the full life-time of an aircraft as long as it does not get into unusual situations.

## NOTES

1. Due to the jet velocity for optimum constant blowing being close to sonic, no comparison was possible with pulsed blowing at the same momentum coefficient. Therefore, the comparison was made with constant blowing at reduced momentum coefficient, equivalent to the one of pulsed blowing with a sonic jet.
2. In [38], eq. (2) the wing span $b$ is not noted. This is due to the pure two-dimensional formulation, where the mass flow is only the sectional mass flow.
3. Spence [41] eq. (36) uses a coefficient $C_J$ for the jet momentum that in contrast to the momentum coefficient is not normalized by the wing area. In the later publication of Maskell & Spence [39] this is corrected.
4. Johnson [54] uses the momentum coefficient symbol $C_\mu$ for the thrust coefficient.
5. Johnson [56], in an earlier publication [54] it was called "flap turning efficiency" as in Figure 6.30.

## REFERENCES

[1] Poisson-Quinton P (1948) Recherches théoriques et expérimentales sur le contrôle de couche limite, 7th Congress of Applied Mechanics, London.
[2] Radespiel R, Burnazzi M, Casper M, Scholz P (2016) Active flow control for high lift with steady blowing, The Aeronautical Journal **120**(1223), pp. 171–200.
[3] Greenblatt D, Wygnanski IJ (2000) The control of flow separation by periodic excitation, Progress in Aerospace Sciences **36**(7), pp. 487–545.
[4] Prandtl L (1904) "Über die Flüssigkeitsbewegung bei sehr kleiner Reibung," III Internationaler Mathematiker-Kongress, Heidelberg, translated to English as: Motion of Fluids with Very Little Viscosity, NACA TM 452 (1928).
[5] Weyl AR (1945) High-lift devices and tailless airplanes, Aircraft Engineering, **17**(10), pp. 292–297.
[6] Betz A, Ackeret J (1928) Quertriebskörper, wie Tragflügel, Schraubenflügel und dergl., German Patent 458,428.
[7] Maillart, G (1947) Aspiration de la couche limite, Bulletin des Services Techniques 106, translation as: Boundary Layer Control by Means of Suction, NASA TM 75502 (1980).
[8] Fage FRS, Sargent RF (1944) Design of suction slots, ARC. R&M 2127.
[9] McCullough GB, Gault DE (1948) An experimental investigation of an NACA $63_1$-012 airfoil with leading edge suction slots, NACA TN 1683.

[10] Glauert MB, Walker WS, Raymer WG, Gregory N (1952) Wind tunnel tests on a thick suction aerofoil with a single slot, ARC. R&M 2646.

[11] Pankhurst RC, Raymer WG, Devereux AN (1953) Wind-tunnel tests of stalling properties of an 8 per cent thick symmetrical section with nose suction through a porous surface, ARC R&M 2666.

[12] Bridges, DH (2007) Early flight-test and other boundary-layer research at Mississippi state 1949–1960, AIAA Journal of Aircraft **44**(5), pp. 1635–1652.

[13] Schlichting H, Gersten K (2006) Grenzschichttheorie, 10th edition, Springer Berlin – Heidelberg, New York.

[14] Schwier W (1947) Versuche zur Auftriebssteigerung durch Ausblasen von Luft an einem Profil von 12% Dicke mit verschiedenen Klappenformen, Deutsche Luftfahrtforschung, Forschungsbericht Nr. 1658, translated into English by Pikler, H (1947) Lift Increase by Blowing Out Air, Tests On Airfoil Of 12 Percent Thickness, Using Various Types Of Flaps, NACA TM 1148.

[15] Gratzer LB, O'Donnell TJ (1965) Development of a BLC high-lift system for high-speed airplanes, Journal of Aircraft **2**(6), pp. 477–484.

[16] Wild J, Chan F, Rudnik R (2008) Active separation control on the leading edge of a slatless swept wing in high lift configuration using steady blowing in the European project EUROLIFT II, KATNet II Separation Control Workshop.

[17] Seifert A, Darabi A, Wygnanski I (1996) Delay of airfoil stall by periodic excitation, Journal of Aircraft **33**(4), pp. 691–698.

[18] Meunier M (2009) Simulation and optimization of flow control strategies for novel high-lift configurations, Journal of Aircraft **47**(5), pp. 1145–1157.

[19] Haucke F, Peltzer I, Nitsche W (2008) Active separation control on a slatless 2D high-lift wing section, 26th International Congress of the Aeronautical Sciences, paper ICAS 2008-3.11.2.

[20] Ciobaca V, Dandois J, Bieler H (2014) A CFD benchmark for flow separation control application, International Journal of Flow Control **6**(3), pp. 67–81.

[21] Meunier M, Dandois (2008) Simulations of novel high-lift configurations equipped with passive and active means of separation control, 4th Flow Control Conference, AIAA paper 2008-4080.

[22] Petz R, Nitsche W (2007) Active separation control on the flap of a two-dimensional generic high-lift configuration, Journal of Aircraft **44**(3), pp. 865–874.

[23] Schatz M, Thiele F, Petz R, Nitsche W (2004) Separation control by periodic excitation and its application to a high lift configuration, 2nd AIAA Flow Control Conference, AIAA Paper 2004-2507.

[24] Johnston JP, Nishi M (1990) Vortex generator jets – means for flow separation control, AIAA Journal **28**(6), pp. 989–994.

[25] Godard G, Stanislav M (2006) Control of a decelerating boundary layer. Part 1: Optimization of passive vortex generators, Aerospace Science and Technology **10**, pp. 181–191.

[26] Ortmanns J, Bitter M, Kähler CJ (2008) Dynamic vortex structures for flow-control applications, Experiments in Fluids **44**, pp. 397–408.

[27] Scholz P, Ortmanns J, Kähler CJ, Radespiel R (2005) Performance optimization of jet actuator arrays for active flow control, KATNet Separation Control Workshop, paper 39.

[28] Wallis RA (1960) A preliminary note on a modified type of air-jet for boundary layer control, Aeronautical Research Council, Australia, Current-Paper CP 513.

[29] Godard G, Foucaut JM, Stanislas M (2006) Control of a decelerating boundary layer. Part 2: Optimization of slotted jets vortex generators, Aerospace Science & Technology **10**(5), pp. 394–400.

[30] Godard G, Stanislas M (2006) Control of a decelerating boundary layer. Part 3: Optimization of round jets vortex generators, Aerospace Science & Technology **10**(6), pp. 455–464.

[31] Warren (1962) Fluidic Oscillator, US Patent 3,016,066.

[32] Graff E, Seele R, Lin JC, Wygnanski I (2013) Sweeping jet actuators – A new design tool for high lift generation, innovative control effectors for military vehicles (avt-215).

[33] Huang LS, Maestrello L, Bryant TD (1987) Separation control over an airfoil at high angles of attack by sound emanating from the surface, AIAA 19th Fluid Dynamics, Plasma Dynamics and Lasers Conference, AIAA Paper 87-1261.

[34] Hsiao FB, Liu CF, Shyu JY (1990) Control of wall-separated flow by internal acoustic excitation, AIAA Journal **28**(8), pp. 1440–1446.

[35] Amitay M, Honohan A, Trautman M, Glezer A (1997) Modification of the aerodynamic characteristics of bluff bodies using fluidic actuators, 28th AIAA Fluid Dynamics Conference, AIAA Paper 97-2004.

[36] Gallas Q, Holman R, Nishida T, Carroll B, Sheplak M, Cattafesta L (2003) Lumped element modeling of piezoelectric-driven synthetic jet actuators, AIAA Journal **41**(2), pp. 240–247.

[37] Liddle SC, Wood NJ (2005) Investigation into clustering of synthetic jet actuators for flow separation control applications, The Aeronautical Journal **109**(1091), pp. 35–44.

[38] Gomes LD, Crowther WJ, Wood NJ (2006) Towards a practical piezoceramic diaphragm based synthetic jet actuator for high subsonic applications – Effect of chamber and orifice depth on actuator peak velocity, 3rd AIAA Flow Control Conference, AIAA Paper 2006-2859.

[39] Lan CE, Campbell JF (1975) Theoretical aerodynamics of upper-surface-blowing jet-wing interaction, NASA TN D-7936.

[40] Dumitrache A, Frunzulica F, Dumitrescu H, Cardos V (2014) Blowing jets as a circulation flow control to enhancement the lift of wing or generated power of wind turbine, INCAS Bulletin **6**(2), pp. 33–49.

[41] Hagedorn H, Ruden P (1938) Windkanaluntersuchungen an einem Junkers- Doppelflügel mit Ausblaseschlitz am Heck des Hauptflügel. Bericht A64 der Lilienthal-Gesellschaft für Luftfahrtforschung.

[42] Spence DA (1956) The lift coefficient of a thin, jet-flapped wing, Proceedings of the Royal Society of London, vol. **238**, pp. 46–68.

[43] Maskell EC, Spence DA (1959) A theory of the jet flap in three dimensions, Proceedings of the Royal Society of London, A Mathematical and Physical Sciences Series **251**(1266), pp. 407–425.

[44] Coandă H (1935) Device for deflecting a stream of elastic fluid projected into an elastic fluid, US patent 2,052,869, originating from Procédé et dispositif pour faire dévier une veine de fluide pénétrant dans un autre fluide, French patent 792,754 (1934).

[45] Coandă H (1961) Jet sustained aircraft, US patent 2,989,303.

[46] Merriam-Webster Dictionary https://www.merriam-webster.com/dictionary/Coanda%20 effect, accessed 2021.

[47] Englar RJ (1975) Experimental investigation of the high velocity Coanda wall jet applied to bluff trailing edge circulation control airfoils, David W. Taylor Naval Ship Research and Development Center Report 4708.

[48] Radespiel R, Pfingsten KC, Jentsch C (2009) Flow analysis of augmented high-lift systems, In: Radespiel R, Rossow CC, Brinkmann BW (Eds) Hermann Schlichting – 100 Years. Notes on Numerical Fluid Mechanics and Multidisciplinary Design, **102**, Springer, Berlin, Heidelberg, pp. 168–189.

[49] Englar RJ (2000) Circulation control pneumatic aerodynamics; blown force and moment augmentation and modification: past, present and future, fluids 20000 conference and exhibit, AIAA Paper 2000-2541.

[50] Englar RJ (1975) Subsonic two-dimensional wind tunnel investigations of the high lift capability of circulation control wing sections, David W. Taylor Naval Ship R&D Center, Report ASED-274.

[51] Pfingsten KC, Cecora RD, Radespiel R (2009) An experimental investigation of a gapless high-lift system using circulation control, KATnet II Conference on Key Aerodynamics Technologies.

[52] Thomas F (1962) Untersuchungen über die Erhöhung des Auftriebes von Tragflügeln mittels Grenzschichtbeeinflussung durch Ausblasen, Zeitschrift für Flugwissenschaften 10(2), pp. 46–65.

[53] ShinMaywa Industries, Ltd. (2020) https://www.shinmaywa.co.jp/aircraft/english/us2/index.html, accessed 2021.

[54] Dandois J, Molton P, Lepage A, Geeraert A, Brunet V, Dor JB, Coustols E (2013) Buffet characterization and control for turbulent wings, AerospaceLab 6, pp. 1–17.

[55] Quigley HC, Innis RC, Holzhauser CA (1964) A flight investigation of the performance, handling qualities, and operational characteristics of a deflected slipstream STOL transport airplane having four interconnected propellers, NASA TN-D-2231.

[56] Beck N, Radespiel R, Lenfers C, Friedrichs J, Rezaeian A (2015) Aerodynamic effects of propeller slipstream on a wing with circulation control, Journal of Aircraft 52(5), pp. 1422–1436.

[57] Burnazzi M, Radespiel R (2014) Assessment of leading-edge devices for stall delay on an airfoil with active circulation control, CEAS Aeronautical Journal 5(4), pp. 359–385.

[58] Johnson WG Jr. (1972) Aerodynamic and performance characteristics of externally blown flap configurations, in: STOL technology, NASA SP 320, pp. 43–54.

[59] Roe H (1973) STOL tactical aircraft investigation, externally blown flap. volume II design compendium, Air Force Flight Dynamics Laboratory report AFFDL-TR-73-20, Vol. 2.

[60] Johnson WG Jr. (1975) Aerodynamic characteristics of a powered, externally blown flap STOL transport model with two engine simulator sizes, NASA TN D-8057.

[61] Heald ER (1973) External blowing flap technology on the Usaf Mcdonnell Douglas yc-15, national aerospace engineering and manufacturing meeting, SAE 1973 Transactions-V82-A, SAE paper 730915.

[62] Tavernetti LR (1992) The C-17: Modern airlift technology, 1992 aerospace design conference, AIAA paper 92–1262.

[63] Yen KT (1982) An analysis of the flow turning characteristics of upper-surface blowing devices for STOL aircraft. Report No. NADC-82007-60.

[64] Riddle DW, Eppel JC (1986) A potential flight evaluation of an upper-surface-blowing/ circulation-control-wing concept, Proceedings of the NASA Circulation-Control Workshop, pp. 539–567 (SEE N88-17586 10-02).

[65] Phelbs, AE III. (197) Aerodynamics of the upper surface blown flap, in: STOL technology, NASA SP 320, pp. 97–110.

[66] Sleeman, WC Jr., Phelps, AE III (1976) Upper surface-blowing flow-turning performance, NASA SP 406, pp. 29–43.

[67] Nark TC (1993) Design, development and flight evaluation of the Boeing YC-14 USB powered lift aircraft, AGARD CP-515.

[68] Shovlin MD, Cochrane JA (1978) An overview of the quiet short-haul research aircraft program, NASA-TM-78545.

[69] Riddle DW, Innis RC, Martin JL, Cochrane JA (1981) Powered-lift takeoff performance characteristics determined from flight test of the quiet short-haul research aircraft (QSRA), AIAA/SETP/SFTE/SAE/ITEA/IEEE 1st Flight Testing Conference, AIAA Paper 81-2409.

[70] Queen S, Cochrane J (1982) Quiet short-haul research aircraft joint navy/NASA sea trials, Journal of Aircraft 19(8), pp. 655–660.

[71] Mori M, Hayashi Y, Takasaki N, Tsujimoto T (1990) Quiet STOL research aircraft ASUKA – development and flight test, ICAS 90-2.7.2.

[72] Sweetman B (1978) New STOL freighter unveiled, flight international, 21st January 1978, p. 163.

[73] Antonov Company (2021) AN-72 light transport airplane, https://www.antonov.com/en/history/an-72, accessed 2021.

[74] Seifert A, Eliahu S, Greenblatt D, Wygnanski I (1998) Use of piezoelectric actuators for airfoil separation control (TN), AIAA Journal **36**(8), pp. 1535–1537.

[75] Riddle DW, Innis RC, Martin JL, Cochrane JA (1981) Powered-lift takeoff performance characteristics determined from flight test of the quiet short-haul research aircraft (QSRA), AIAA/SETP/SFTE/SAE/ITEA/IEEE 1st Flight Testing Conference, AIAA Paper 81-2409.

[76] Keller D (2013) Numerical investigation of engine effects on the stability and controllability of a transport aircraft with circulation control, 31st AIAA Applied Aerodynamics Conference, AIAA paper 2013-3031.

[77] Loth JL (2005) Why have only two circulation-controlled STOL aircraft been built and flown in years 1974–2004. Proceedings of the 2004 NASA/ONR Circulation Control Workshop, NASA CP-2005-213509, pp. 603–640.

# 7 Simulation of High-Lift Flows

## NOMENCLATURE

| | | | | | |
|---|---|---|---|---|---|
| $a$ | $m/s$ | Speed of sound | $T$ | $K$ | Temperature |
| $A$ | $m^2$ | Effective cross-sectional area | $T$ | – | Tunnel shape factor |
| $A_{ref}$ | $m^2$ | Reference area | $\mathbf{u}$ | $m/s$ | Velocity vector |
| $A_0$ | $m^2$ | Actual cross-sectional area | $u,v,w$ | $m/s$ | Velocity components |
| $b$ | $m$ | Span | $U$ | $m/s$ | Flow velocity |
| $B$ | $m$ | Test section width | $V$ | $m^3$ | Volume |
| $c$ | $m$ | Chord length | $\mathbf{x}$ | $m$ | Coordinate vector |
| $c_D$ | – | Section drag coefficient | $x,y,z$ | $m$ | Cartesian coordinates |
| $c_L$ | – | Section lift coefficient | $\alpha$ | ° | Angle of attack |
| $c_p$ | – | Pressure coefficient | $\alpha_{eff}$ | ° | Effective angle of attack |
| $c_{ref}$ | $m$ | Reference chord length | $\alpha_{SC}$ | ° | Angle of attack due to streamline curvature |
| $C$ | $K$ | Sutherland constant | $\beta$ | – | Prandtl-Glauert factor |
| $C$ | $m^2$ | Test section cross-sectional area | $\Gamma$ | $m^2/s$ | Vortex strength |
| $C_D$ | – | Drag coefficient | $\delta$ | $m$ | Bedding displacement |
| $C_L$ | – | Lift coefficient | $\delta$ | – | Vertical flow induction coefficient |
| $C_m$ | – | Pitching moment coefficient | $\delta_\varepsilon$ | – | Vertical flow induction coefficient due to blockage |
| $D$ | $N$ | Drag force | $\delta_2$ | $m$ | Momentum loss thickness |
| $E$ | $N/m^2$ | Young's modulus | $\varepsilon$ | – | Axial flow induction coefficient |
| $E$ | $J$ | Inner energy | $\varepsilon$ | ° | Local twist angle |
| $f$ | $Hz$ | Frequency | $\varepsilon_w$ | – | Axial flow induction coefficient due to wake blockage |
| $F$ | $N$ | Force | $\varepsilon_\delta$ | – | Axial flow induction coefficient due to lift |
| $G$ | $N/m^2$ | Shear modulus | $\varepsilon_0$ | – | Axial flow induction coefficient at model position |
| $H$ | $m$ | Test section height | $\theta$ | ° | Angle |
| $H$ | $J$ | Enthalpy | $\kappa$ | – | Isentropic exponent |
| $l$ | $m$ | Length | $\lambda$ | – | Body shape factor |
| $L$ | $N$ | Lift force | $\lambda_T$ | $W/m \cdot K$ | Heat conductivity |
| $L$ | $m$ | Test section length | $\Lambda$ | – | Wing aspect ratio |
| $I_{ii}$ | $m^4$ | Second area moment of inertia | $\mu$ | $m^2/s$ | Dipole strength |
| $I_p$ | $m^4$ | Polar area moment of inertia | $\mu$ | $kg/m^2$ | Dynamic viscosity |

DOI: 10.1201/9781003220459-7

| | | | | | |
|---|---|---|---|---|---|
| $M$ | – | Mach number | $\xi, \zeta$ | – | Dimensionless tunnel coordinates |
| $M$ | $Nm$ | Moment | $\rho$ | $kg/m^3$ | Density |
| $M_b$ | $Nm$ | Bending moment | $\tau$ | $N/m^2$ | Shear stress tensor |
| $M_t$ | $Nm$ | Torsion moment | $\varphi$ | $^\circ$ | Torsion twist angle |
| $M_y$ | $Nm$ | Pitching moment | $\phi$ | – | Fluid property |
| $\mathbf{n}$ | $m^2$ | Surface normal vector | $\Phi$ | $m^2/s$ | Potential |
| $p$ | $N/m^2$ | (Static) pressure | $\Phi_w$ | $m^2/s$ | Wall interference potential |
| $q$ | $N/m^2$ | Dynamic pressure | $\Omega$ | $m^2$ | Surface |
| $q_i$ | $m^2/s$ | Source strength | $_c$ | | Corrected |
| $r$ | $m$ | Radius | $_i$ | | Induced |
| $R$ | $J/kg \cdot K$ | Universal gas constant | $_t$ | | Stagnation state |
| $Re$ | – | Reynolds number | $_u$ | | Uncorrected |
| $s$ | $m$ | Arc length along body surface | $_0$ | | Reference state |
| $S$ | – | Strouhal number | $_{2D}$ | | Two-dimensional |
| $t$ | $m$ | Thickness | $_\infty$ | | Onflow condition |
| $t$ | $s$ | Time | | | |

Other than earthbound transport vehicles, an aircraft has to be safely able to fly from the beginning of its maiden flight. It is, therefore, not possible to completely build a prototype and test it without a detailed a priori knowledge of its aerodynamic and flight-mechanical behavior. In consequence, it is necessary to gain knowledge beforehand, which is done by simulating the aerodynamic behavior. In general, such simulations can be performed experimentally or numerically.

Due to the nature of the mathematics describing the flow, in the beginning, it was not possible to predict the full aerodynamics of an aircraft by calculations. So, from the beginning of aviation, the experimental simulation was the major way to gain insight into the flight characteristics of an aircraft. With upcoming computer technology, numerical simulations gained importance. In the 1970s, it was possible to simulate simplified aircraft flow by potential theory, sometimes coupled to boundary layer to consider viscous effects. In the 1980s, it became possible to compute the flow around an aircraft in transonic conditions based on the Euler equations, and, in the 1990s, the computer resources have got large enough to even compute on the basis of the Navier-Stokes equations.

In this time, there was rumor that numerical simulation may completely substitute experimental simulation and there was rivalry between experiments and numerics. Nowadays, a common understanding is that the best way for simulation is a clever combination of both using the strengths and avoiding or bridging the weakness of the one simulation stream by the other.

## 7.1   EXPERIMENTAL SIMULATION

As for a large transport aircraft, it is not possible to build a full-size mock-up and to expose it to the real air flow, experimental simulation is mostly made on smaller scale models. From relativity theory, it is mechanically not important whether the

aircraft is flying with its own speed through still air, or the model rests in an air flow
of the same velocity. It is only a question of the reference frame and the observer
position. For the observation of the flow and for routing of measurement transmis-
sion lines, it is better to keep the observer at the aircraft as it is easier to implement.

The smaller scale of the model imposes the problem that in air a fully similar
flow in comparison to the real aircraft cannot be obtained. Dimensional analysis of
the governing Navier-Stokes equations shows that some characteristic numbers have
to be equal for this similarity. For steady flows around aircraft, these are mainly the
Reynolds number and the Mach number.

The Reynolds number

$$Re = \frac{\rho \cdot l \cdot U}{\mu} \qquad (7.1)$$

characterizes the ratio of the inertial forces to the viscous forces. The higher the
Reynolds number the less is the impact of friction. The Mach number

$$M = \frac{U}{a} \qquad (7.2)$$

describes the compressibility of the flow by relating the velocity to the speed of
sound. Remembering the discussion on the role of compressibility for the limits of
lift generation in Section 2.2, the Mach number is even as important as the Reynolds
number for the simulation of high-lift flows.

While the Mach number is not scale-dependent and can be easily achieved by
selecting the flow speed, the Reynolds number is dependent on velocity and scale.
It is, therefore, not possible to achieve, with a scaled aircraft model, the same Mach
and Reynolds number as the real aircraft at the same time without changing the fluid
properties. The speed of sound of a perfect gas is directly linked to the temperature,
and the Mach number expressed with the temperature is

$$M = \frac{U}{\sqrt{\kappa \cdot R \cdot T}}. \qquad (7.3)$$

Alternatively, the only way to change the Mach number related to a certain flow
speed is to change the fluid to something not close to the perfect gas, e.g., water. For
the Reynolds number, the dynamic viscosity also depends on temperature by the
Sutherland law

$$\mu(T) = \mu_0 \frac{T_0 + C}{T + C} \left( \frac{T}{T_0} \right)^{3/2} \qquad (7.4)$$

with the Sutherland constant $C = 110\ K$ for a perfect gas. The density varies depending on temperature and pressure according to the universal gas law

$$\frac{p}{\rho} = RT. \qquad (7.5)$$

The Reynolds number can therefore also be expressed as a function of Mach number, pressure, and temperature

$$Re = \frac{\kappa \cdot p \cdot l}{\mu(T)} \frac{M}{\sqrt{\kappa \cdot R \cdot T}}. \tag{7.6}$$

The Reynolds number at a certain Mach number can therefore be increased by increasing the pressure and/or reducing the temperature.

For unsteady flows, additionally the Strouhal number

$$S = \frac{f \cdot l}{U} \tag{7.7}$$

is of importance. It relates the time scale of the unsteady flow field to the convective time scale of the flow, namely the time the flow needs to travel at the given velocity $U$ the reference length $l$. Also, the Strouhal number is therefore scale-dependent, and the unsteadiness imposed on the flow must be faster in model scale.

### 7.1.1 WIND TUNNELS

Wind tunnels are experimental facilities to provide an air flow to a model of the aerodynamic body mounted inside. There are some basic characteristics that differentiate the wind tunnels into different categories.

The first characteristic is the type of circuit. In general, three types of wind tunnel circuits are widely used, depicted in Figure 7.1. The first two types are used for continuously providing the air flow. They are both driven by a fan. The Eiffel[1]-type wind tunnel (Figure 7.1(a)) has an open circuit. The driving fan is at the end of the tunnel, so outside ambient air is sucked through the tunnel. It is, therefore, not possible to control the fluid state of the flow. In a closed wind tunnel circuit of the Göttingen[2]-type (Figure 7.1(b)), the flow is circulated inside a closed channel. It is, therefore, possible to change the fluid state in this closed volume. Additionally, the power needed to drive the wind tunnel is only due to the viscous and pressure losses in the wind tunnel circuit. For an open circuit, the power needed is for the full acceleration of the air flow as the incoming flow is at rest beforehand.

The last type of tunnel circuits provides the air flow only for a distinct amount of time. Such a blow down tunnel (Figure 7.1(c)) consists of a pressurized air reservoir and an evacuated low-pressure reservoir. When opening control valves, air flows for a certain amount of time, depending on the reservoir size and the designated flow speed. Such tunnel types are mainly used for very high flow speeds in the supersonic and hypersonic flow regimes.

The second characteristic to differentiate wind tunnels is the type of test section. Figure 7.2 illustrates the two most common options. In open test sections, the circuit is interrupted at the test section where the model is placed. Closed test sections provide a well-defined state of the air flow. "The boundary condition for a closed tunnel is exact and precise, except for any effects due to the frictional boundary layer along the walls" – (Glauert [1] p. 3). Nevertheless, in comparison to free air flow, the

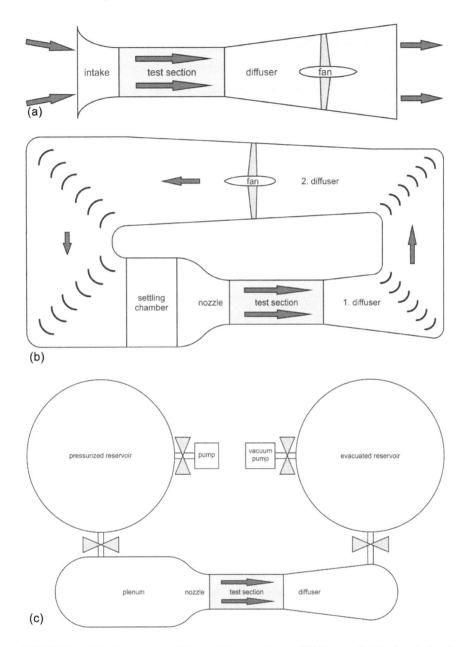

**FIGURE 7.1**  Wind tunnel circuit types: (a) open circuit (Eiffel-tunnel); (b) closed circuit (Göttingen-tunnel); (c) blow-down.

presence of the wall imposes a blockage of the air flow, especially when large wind tunnel models are used. This impact will be discussed in Section 7.1.3. To mitigate the influence of the walls, some wind-tunnels implement slotted walls or adaptive walls. "The principal objective of the ventilated walls is to allow the stream to cross

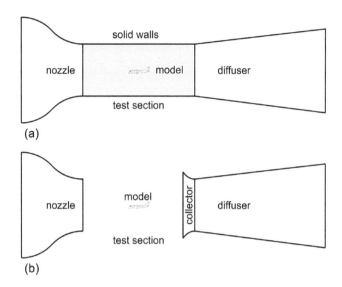

**FIGURE 7.2**    Wind tunnel test section types: (a) closed test section; (b) open test section.

the test section boundary, to generate a flow pattern resembling the one in uncon-strained flow. Airflow through the ventilated wall (perforated or slotted) is regulated by adjusting the pressures in segmented plenum compartments [...] and by varia-tions of the open area ratio. With these mainly the acceleration of the flow due to the blockage of the model shall be minimized" – (Mokry et al. [2] p. 167).

Open test sections especially provide a good optical access to the model for observation of the flow field. "It is important to point out that an open jet bound-ary differs from free air because in the latter case the pressure perturbations due to the presence of the model vanish, in general, only at infinity. The curvature of the open jet boundary is greater than that of a corresponding streamline in infinite stream, since there is no outside flow to resist the deformation. A flow pattern will be obtained in which the streamlines are further apart than for free flight" – (Mokry et al. [2] p. 5).

Further characteristics of the tunnel type are the designated flow speed (subsonic, transonic, supersonic, hypersonic), the fluid used (air, nitrogen, water), and the fluid state (ambient, pressurized, cryogenic). As shown in eq. (7.6), the Reynolds number can be increased by reducing the temperature and increasing the pressure. A pressur-ized low-speed wind tunnel is, e.g., the ONERA F1 wind tunnel in Fauga-Mauzac, France [3], which can be pressurized up to $p = 3.85\,\text{bar}$, and the Low Turbulence Pressure Tunnel (LTPT) at NASA Langley, Hampton, Virginia, US [4] with static pressures up to $p = 10\,\text{atm}\left(10.1325\,\text{bar}\right)$. The use of nitrogen as fluid allows for cryo-genic conditions, namely very low temperatures, since the boiling temperature of nitrogen is at about $T = 90\,\text{K}$ at ambient pressure. Wind tunnels that apply cryogenic conditions are the Cryogenic Wind Tunnel Cologne (DNW-KKK), Germany [5], the European Transonic Wind Tunnel (ETW) in Cologne, Germany [6], and the National Transonic Facility (NTF) at NASA Langley, Hampton, Virginia, USA [7].

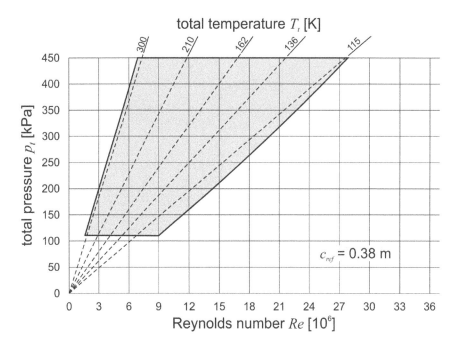

**FIGURE 7.3** Reynolds number in cryogenic conditions for a typical high-lift configuration model depending on temperature and pressure. (From Germain & Quest [8], figure 2 right.)

In the last two, the tunnel can additionally be pressurized to obtain even higher Reynolds numbers.

Figure 7.3 shows the dependency of the Reynolds number on temperature and pressure for a typical high-lift aircraft model in the ETW. The ambient condition is the lower left corner of the envelope. From eq. (7.6), the linear scaling of the Reynolds number with the pressure is obvious. The Reynolds number can further be increased by a factor of 4 by the cryogenic environment. Nevertheless, during measurements, it is important to respect the local velocity. Due to the cooling of accelerated flow, the nitrogen may turn into liquid state resulting in condensation on the model.

## 7.1.2 WIND TUNNEL MODEL TYPES AND INSTALLATIONS

Depending on the nature of investigation, different types of wind tunnels are used for investigations of high-lift flows. In case of studies on the aerodynamics of the high-lift airfoil, two-dimensional models are used. They consist of a constant airfoil section at constant chord along their span. Mostly, such two-dimensional airfoil models are mounted from one wall to the other (Figure 7.4 (a)). "If [...] a wing stretches from wall to wall of a closed rectangular tunnel there will be no system of trailing vortices, apart from any minor effects due to the boundary layer of the tunnel [...]" – (Glauert [1] p. 41). If the model is of less span, it may be mounted with side plates (Figure 7.4 (b)) or as cantilever wing (Figure 7.4 (c)), the latter also allowing for swept

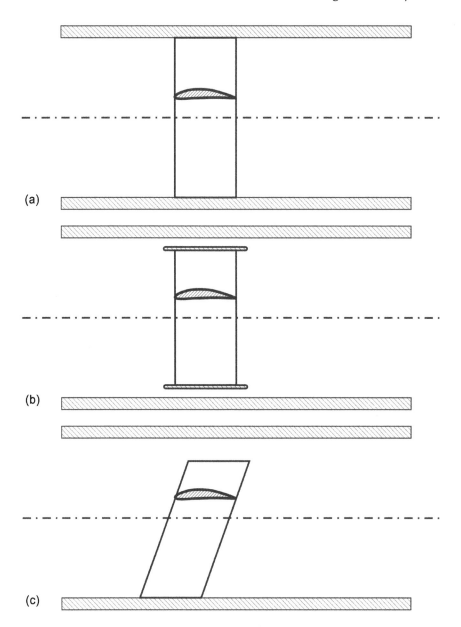

(a)

(b)

(c)

**FIGURE 7.4**   Model types and wind tunnel installations of airfoil models: (a) wall-to-wall; (b) side plates; (c) cantilever wing.

wing installation. In theory, the quasi two-dimensional airfoil characteristics are then obtained for infinite span only. It is, therefore, worth to implement an aspect ratio

$$\Lambda_{2D} = \frac{b^2}{A_{ref}} = \frac{b}{c} \tag{7.8}$$

as high as possible. On the other hand, the larger the chord length, the higher the Reynolds number obtained in the wind tunnel. At the end, the aspect ratio of a two-dimensional model has to be a compromise between obtaining mostly proper two-dimensional flow and high Reynolds numbers at the same time. In order to measure mostly two-dimensional airfoil aerodynamics in the most part of the center of the model, the aspect ratio of such a model should be around $\Lambda_{2D} = 4$.

For the investigation of the aerodynamics of high-lift wings and aircraft in high-lift configuration, the models used are a representative down-scale of the aircraft with all relevant details. The empennage is often omitted as it does not significantly affect the high-lift wing flow. But, supporting structures such as flap and slat tracks are often tried to be included by their respective disturbance of the flow.

Aircraft models can be either built as full models (Figure 7.5 (a)) mounted on a sting or support structure or as half models mounted on the wind tunnel wall (Figure 7.5 (b)). While the full model is not directly influenced by the wind tunnel

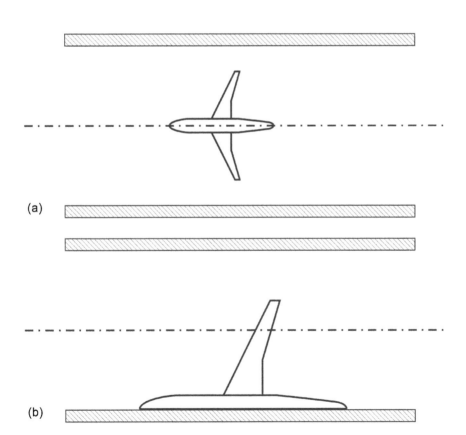

**FIGURE 7.5** Model types and wind tunnel installations of aircraft models: (a) full model; (b) half model.

wall flow, the Reynolds number achievable with half models is, in principle, doubled. According to Franz [9]:

> The most important advantages [of half-models for high-lift testing] are:
> - for a given wind tunnel the possible model scale is about twice as great as for the comparable full-scale model. This means a doubling of the Re-number for flow mechanics and thus a strong approximation of the flight Re-numbers,
> - in the model fabrication the mirror-symmetrical half is eliminated which reduces costs.
> - the largest half-model parts can be fabricated more simply as a rule and, for equal care, with double the accuracy of the full-model,
> - the adjustment accuracy of the individual elements with respect to one another is increased considerably. This is of great importance especially with regards to the positioning of individual flaps for highly developed multiple-flap system.
>
> **(Franz [9] p. 7)**

### 7.1.3    WIND TUNNEL CORRECTIONS

The aerodynamic situation of an air stream inside the wind tunnel differs substantially from the conditions of a free air flow. Put simply, the mass of air affected by the installed model is limited by the cross-section of the wind tunnel air stream, be it a closed or an open test section. "The nature of tunnel-wall constraint can be deduced from physical principles of streamline flow. It is also associated directly with the theoretical consideration that, although the differential equations of the flow are the same in the tunnel as in free air, the outer boundary conditions are different" – (Garner [10] p. 5).

It has been a major topic since the beginning of wind tunnel testing to try to find theoretical relationships to quantify the deviations of the measured values from the situation in free air flow. For some of the systematic deviations, analytic correlations have been deduced that can be applied directly. The two AGARDographs AG 109 [10] and AG 336 [11] provide a collection of state-of-the-art wind tunnel corrections. For other types of errors, it is not able to address them in the same way and they have to be minimized as far as possible. Those are discussed in the last part of this section.

For the analytical derivation of wind tunnel corrections, first, it is valuable to define non-dimensional characteristics of the deviations introduced by the limited airflow. An induced deviation of the axial flow velocity can be described as

$$\varepsilon = \frac{u_i}{U_\infty}. \tag{7.9}$$

For the vertical component, it is common to interpret the deviation as a change of effective angle of attack.

$$\Delta\alpha = \frac{w_i}{U_\infty}. \tag{7.10}$$

As the induction of vertical components is mostly – but not exclusively – related to the generation of lift, a non-dimensional upwash interference factor is given by

$$\delta = \frac{w_i}{U_\infty} \frac{C}{A_{ref} \cdot C_L} \tag{7.11}$$

with its derivative regarding along the test section length

$$\delta_1 = \frac{\partial \delta}{\partial \left( \frac{x}{\beta L} \right)} \tag{7.12}$$

"$C$ is the test section cross-sectional area, and $L$ is a typical length scale (often taken as the height of the test section $H$)" – (Krynytzky in [11] p. 2-8).

Eq. (7.12) includes the Prandtl-Glauert factor

$$\beta = \sqrt{1 - M_\infty^2} \tag{7.13}$$

to account for compressibility effects. The space coordinates are transferred to dimensionless coordinates by referring to the height of the wind tunnel

$$\xi = \frac{x}{\beta H}$$
$$\zeta = \frac{z}{H}. \tag{7.14}$$

The principal method for wind tunnel corrections applies potential theory [1, 12], assuming that the model is small enough to be represented as a single elementary singularity. To account for the existence of wind tunnel walls, a similar procedure as for the ground effect described in Section 4.1.2 by mirroring the elementary singularity is applied. In order to establish the two walls on each side of the model, the mirroring has to be performed on both sides – the original one and the mirrored one – leading to an infinite series of singularities as depicted schematically in Figure 7.6 for the two-dimensional situation. In three dimensions, the principle is the same and the singularities are additionally mirrored in the second direction perpendicular to the air flow. The interference by the wind tunnel walls is then the flow field induced by all elementary singularities except the one representing the model in the test section.

### 7.1.3.1 Corrections for Lift Interference

The first correction to be applied relates to the interference of the lift generating body with the wind tunnel walls. Figure 7.7 illustrated the distortion of the flow field due to the existence of the wind tunnel walls. Compared to the situation in free stream (dashed streamlines) the wind tunnel walls bend the streamlines before and after the model to reduce the incoming angle of attack and to block the downwash

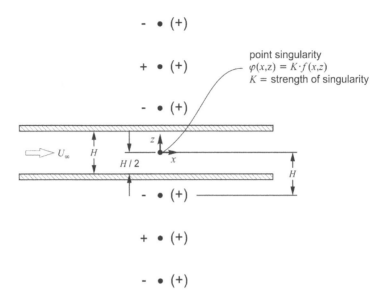

**FIGURE 7.6** Image system for a singularity at the center of a 2D tunnel with closed walls. (According to Krynytzky in [11], figure 2.6.)

**-, +: Image strength for Φ odd in z**
**(+): Image strength for Φ even in z)**

past the lifting body. To represent this, in the singularity arrangement sketched in Figure 7.6, the elementary singularity is taken to be an elementary potential vortex with strength Γ, which is by Zhukovsky's theorem (eq. (2.21) in Chapter 2) proportional to the lift coefficient.

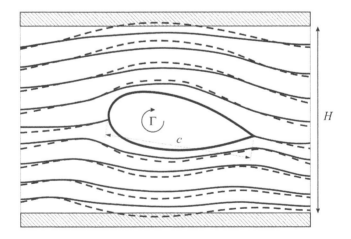

**FIGURE 7.7** Effect of walls on circulation on flow field in a closed test section.

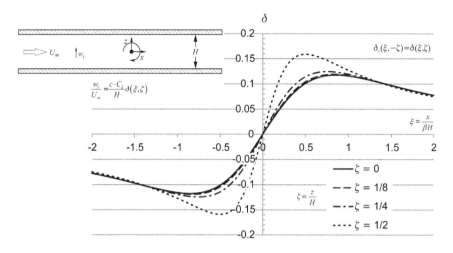

**FIGURE 7.8** Upwash interference of a 2D vortex in a closed-wall tunnel. (According to Krynytzky in [11], figure 2.9.)

The upwash interference factor, shown in Figure 7.8 depending on the lateral position for different vertical displacements out of axis, results in a two-dimensional flow to[3]

$$\delta(\xi,\zeta) = \frac{H}{U_\infty \cdot c \cdot C_L}\frac{\partial \Phi_w}{\partial z} = -\frac{1}{2\pi}\sum_{\substack{n=-\infty \\ n\neq 0}}^{n=\infty}(-1)^n \frac{\xi}{\xi^2 + (\zeta-n)^2}. \tag{7.15}$$

The streamwise interference parameter, depicted in Figure 7.9, is given by[4]

$$\varepsilon_\delta(\xi,\zeta) = \frac{1}{U_\infty}\frac{\partial \Phi_w}{\partial x} = \frac{C_L}{2\pi\beta}\frac{c}{H}\sum_{\substack{n=-\infty \\ n\neq 0}}^{n=\infty}(-1)^n \frac{\zeta-n}{\xi^2 + (\zeta-n)^2}. \tag{7.16}$$

Due to symmetry, the upwash and the axial interference factors are identically zero at the position of the model. Nevertheless, the upwash gradient is not

$$\delta_1(0,0) = \frac{\pi}{24}. \tag{7.17}$$

"Since the upwash gradient is proportional to $C_L$, the uncorrected lift curve will be steeper" – (Krynytzky in [11] p. 2-15). The explanation of the lift changes is properly explained by Krynytzky [11] as: "Although these results are strictly applicable only to a small model, the implications of finite model size are apparent from consideration of the spatial variations of interference velocities in [Figure 7.8] and [Figure 7.9]. A model centred between the walls at zero incidence may have a chord

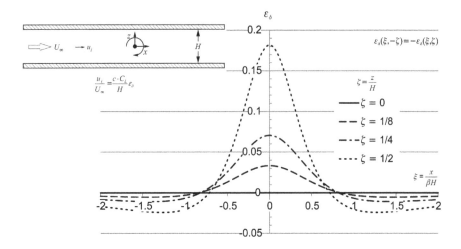

**FIGURE 7.9** Streamwise interference of a 2D vortex in a closed-wall tunnel. (According to Krynytzky in [11], figure 2.10.)

length that places leading and trailing edges beyond the region of 'constant' interference. Further, rotating such a model through a range of incidence angles moves both leading and trailing edges away from the centreline and into regions of variable upwash and streamwise interference. The limits of linear streamwise upwash along the centreline are no more than about $x/\beta H \leq \pm 0.4$. Deviations of both upwash and streamwise interference from the centreline value are small for $z/H \leq \pm 0.2$" – (Krynytzky in [11] p. 2-15).

"In addition to the interference effects associated with airfoil thickness and camber, it is necessary to consider a further alteration of the field of flow caused by the confining influence of the tunnel walls upon the airfoil wake" – (Allen & Vincenti [13] p. 156). Goldstein [14] gives corrections for lift-induced corrections based on streamline curvature. Allen & Vincenti [13] refine this approach by adding compressibility. Those corrections obtained are summarized in [11] as[5]

$$\Delta\alpha = \frac{\pi c^2}{96\beta H^2}(C_L + 4C_m)$$

$$\Delta C_L = -\frac{\pi^2}{48}\left(\frac{c}{\beta H}\right)^2 C_L \qquad (7.18)$$

$$\Delta C_m = -\frac{\pi^2}{192}\left(\frac{c}{\beta H}\right)^2 C_L.$$

As pointed out by Garner [10] "terms in $(c/H)^4$ are omitted. It should therefore be recognized that equations (7.18) are subject to significant inaccuracy where $c > 0.4\beta H$" – (Garner [10] p. 34). For larger models or higher Mach numbers, this limitation gets significant. Havelock [15] reports the corrections including the higher order terms

$$\Delta\alpha = \beta C_L \left\{ \frac{\pi}{96}\left(\frac{c}{\beta H}\right)^2 - \frac{41\pi^3}{92160}\left(\frac{c}{\beta H}\right)^4 + O\left(\frac{c}{\beta H}\right)^6 \right\}$$

$$\Delta C_L = C_L \left\{ -\frac{\pi^2}{48}\left(\frac{c}{\beta H}\right)^2 + \frac{7\pi^4}{3072}\left(\frac{c}{\beta H}\right)^4 + O\left(\frac{c}{\beta H}\right)^6 \right\} \qquad (7.19)$$

$$\Delta C_m = C_L \left\{ \frac{\pi^2}{192}\left(\frac{c}{\beta H}\right)^2 + \frac{7\pi^4}{15360}\left(\frac{c}{\beta H}\right)^4 + O\left(\frac{c}{\beta H}\right)^6 \right\}.$$

Alternatively, if the moment coefficient is measured, the angle of attack correction can be evaluated according to Garner [10] as

$$\Delta\alpha = \frac{\pi c^2}{96\beta H^2}(C_L + 4C_m) - \frac{7\pi^3 c^4 C_L}{30720\beta^3 H^4}. \qquad (7.20)$$

The methodology to obtain wind tunnel corrections for three-dimensional full models is similar. Instead of a singularity in a two-dimensional flow, the model is represented by a lifting line to represent the wing and the trailing vortices. The wind tunnel walls are again respected by mirroring, but now in the two dimensions perpendicular to the wind tunnel axis. Similar analysis leads to the upwash influence factor at the model position

$$\delta(0,0,0) = \frac{B \cdot H}{8\pi} \sum_{m=-\infty}^{m=\infty} \sum_{\substack{n=-\infty \\ n \neq m=0}}^{n=\infty} (-1)^n \frac{m^2 B^2 - n^2 H^2}{\left(m^2 B^2 + n^2 H^2\right)^2} \qquad (7.21)$$

and its derivative

$$\delta_1(0,0,0) = \frac{B \cdot H^2}{8\pi} \sum_{m=-\infty}^{m=\infty} \sum_{\substack{n=-\infty \\ n \neq m=0}}^{n=\infty} (-1)^n \frac{m^2 B^2 - 2n^2 H^2}{\left(m^2 B^2 + n^2 H^2\right)^{5/2}}. \qquad (7.22)$$

"Additional upwash at the model location due to the walls requires corrections to angle of attack and drag [...]" – (Krynytzky in [11] p. 2-21). This accounts for the induced wind axis being not parallel to the wind tunnel axis. Therefore, the forces measured in the wind tunnel axis are not exactly represented in the aerodynamic coordinate system. "For a small model and small upwash angle, the corrections to lift and drag [...] are" – (Krynytzky in [11] p. 2-21)

$$C_{L,c} = C_{L,u} \cos \Delta\alpha(0,0) - C_{D,u} \sin \Delta\alpha(0,0)$$
$$C_{D,c} = C_{D,u} \cos \Delta\alpha(0,0) - C_{L,u} \sin \Delta\alpha(0,0) \qquad (7.23)$$
$$\Delta\alpha(0,0) = \delta(0,0)\frac{A_{ref}}{C}C_{L,u}$$

"Though the above relationships define a corrected onset stream direction, the model angle of attack must additionally be adjusted for interference stream curvature. Because the wing is immersed in an interference flow field characterised by increasing upwash with x, it appears to have an increased effective camber (in a closed-wall tunnel) compared to an unbounded flow" – (Krynytzky in [11] p. 2-21). The additional effect on the onflow direction induced by the streamline curvature is scaling with the derivative of the upwash in streamwise direction

$$\alpha_c = \alpha_u + \Delta\alpha + \Delta\alpha_{SC} = \alpha_u + \left( \delta(0,0) + \frac{c}{2\beta H} \delta_1(0,0) \right) \frac{A_{ref} \cdot C_{L,u}}{C}. \qquad (7.24)$$

The according pitching moment correction is

$$\Delta C_m = \delta_1(0,0) \frac{c}{16\beta H} \frac{A_{ref}}{C} C_{L,u} \frac{\partial C_L}{\partial \alpha} \qquad (7.25)$$

### 7.1.3.2   Corrections for Blockage

The effect of blockage correlates with the displacement of the flow due to the thickness of the model. As sketched in Figure 7.10, due to the existence of the model within the closed wind tunnel test section, the cross-sectional area of the flow is reduced leading to an acceleration of the flow.

For the derivation of blockage corrections of a two-dimensional model, the singularity for the methodology sketched in Figure 7.6 is a dipole of strength

$$\mu = \frac{A \cdot U_\infty}{\beta}, \qquad (7.26)$$

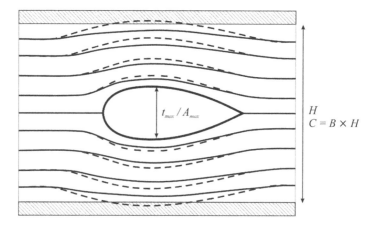

**FIGURE 7.10**   Effect of model blockage on flow field in a closed test section.

where $A$ is the effective cross-sectional area of the model derived as the maximum airfoil thickness. The streamwise interference factor results in

$$\varepsilon(\xi,\zeta)-\frac{1}{2\pi\beta^3}\frac{A}{c^2}\left(\frac{c}{H}\right)^2\sum_{\substack{n=-\infty\\n\neq0}}^{n=\infty}\frac{\xi^2-(\zeta-n)^2}{\left[\xi^2+(\zeta-n)^2\right]^2}. \tag{7.27}$$

At the position of the model, the streamwise interference is

$$\varepsilon(0,0)=\frac{\pi}{6}\frac{A}{\beta^3 H^2}. \tag{7.28}$$

"It should be noted that at any value of $y$, the interference is a maximum at the model location, [Figure 7.11], which increases the effective freestream velocity felt by the model. However, due to the streamwise symmetry of the interference, there is no pressure buoyancy force on the model" – (Krynytzky in [11] p. 2-23).

Analogously to the lift interference, the upwash interference parameter for a non-lifting body is derived to[6]

$$\delta_\varepsilon(\xi,\zeta)=\frac{1}{2\pi\beta^2}\frac{A}{H^2}\sum_{\substack{n=-\infty\\n\neq0}}^{n=\infty}\frac{2\xi(\zeta-n)}{\left[\xi^2+(\zeta-n)^2\right]^2}. \tag{7.29}$$

"By symmetry, the interference upwash due to solid blockage is zero along the axis of the tunnel, [Figure 7.12]. Off-centreline the interference upwash has a character similar to the upwash interference of a 2D vortex" – (Krynytzky in [11] p. 2-24).

As the representation of a two-dimensional airfoil model with a single dipole source is acceptable only for small airfoil chords, a better correction for larger

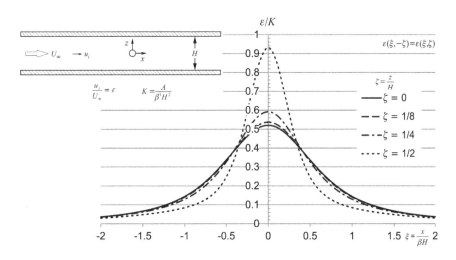

**FIGURE 7.11** Streamwise interference of a 2D source doublet in a closed-wall tunnel. (According to Krynytzky in [11], figure 2.15.)

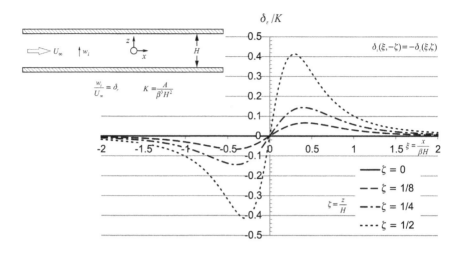

**FIGURE 7.12** Upwash interference of a 2D source doublet in a closed-wall tunnel. (According to Krynytzky in [11], figure 2.16.)

models is obtained by introducing a shape factor that relates the disturbance of an arbitrary 2D body to the disturbance of an equivalent cylinder [1] using the so-called body shape factor, $\lambda$

$$r = 2t_{max}\sqrt{\lambda} \qquad (7.30)$$

and the corresponding dipole strength is derived as

$$\mu = \frac{\pi}{2}\lambda \cdot t_{max}^2 \cdot U_\infty. \qquad (7.31)$$

In general, the body shape factor is obtained for a symmetrical body by the integration of the inviscid velocity distribution

$$\lambda = \frac{4}{\pi}\int_{LE}^{TE} \frac{U(s)}{U_\infty} \frac{z(s)}{t_{max}^2}ds. \qquad (7.32)$$

For cases, where the velocity distribution is not a priori available, Glauert [1] provides an estimate based on the maximum thickness

$$\lambda = \frac{2A_0}{\pi t_{max}^2}, \qquad (7.33)$$

where the effective cross-sectional area $A$ is related to the actual cross-sectional area $A_0$ by

$$A = \left(1 + \frac{t_{max}}{c}\right)A_0. \qquad (7.34)$$

The streamwise interference factor at the model position is deduced as

$$\varepsilon(0,0) = \frac{\pi^2}{12} \frac{\lambda t_{max}^2}{\beta^3 H^2}. \tag{7.35}$$

An alternative relation for the streamwise interference factor is recommended by Rogers [10]

$$\varepsilon(0,0) = \frac{\pi}{6}\left(1 + 1.2\beta \frac{t_{max}}{c}\right)\frac{A}{\beta^3 H^2}. \tag{7.36}$$

An extension for considering the changed influence due to asymmetry under angle of attack is provided by Batchelor [16]

$$\varepsilon(0,0)(\alpha) = \varepsilon(0,0)\left(1 + 1.1\frac{c}{t_{max}}\alpha^2\right). \tag{7.37}$$

A similar analysis for a three-dimensional model is obtained by replacing the model by an equivalent sphere. "For the purpose of calculating the solid-blockage effects the small wing or wing-body combination of finite span may be regarded as identical to a body of revolution of the same volume. This implies, however, that the wing is replaced by an equivalent sphere or spheroid, and not, as one might feel is more appropriate, by an equivalent cylinder" – (Rogers [10] p. 297). The streamwise interference according to Rogers [10] is given by

$$\varepsilon(0,0,0) = T\left(\frac{1}{BH}\right)^{3/2}\frac{V}{\beta^3}\left[1 + 1.2\beta \frac{t_{max}}{c}\right] \tag{7.38}$$

using the tunnel shape factor

$$T = \frac{1}{4\pi}\sum_{m=-\infty}^{m=\infty}\sum_{\substack{m=-\infty \\ n \neq m = 0}}^{n=\infty}\left(\frac{BH}{m^2 B^2 + n^2 H^2}\right)^{3/2} \tag{7.39}$$

in order to respect the impact of the aspect ratio of the test cross-section.

The major consequence of the streamwise interference due to blockage is an increased flow speed at the model position. The wind tunnel instrumentation allows to measure pressures in the tunnel, mainly the stagnation pressure in the settling chamber upstream of the nozzle and the static pressure ratio at the nozzle. The wind tunnel flow speed is obtained from these measurements by calibration. Due to the blockage, the flow velocity obtained from wind tunnel calibration is lower than seen by the model. As the stagnation pressure is unchanged, also the static pressure seen at the model is less than for the empty test section. Therefore, the streamwise interference affects a correction of all reference values used to derive the aerodynamic properties.

$$U_{\infty,c} = U_{\infty,u}\left(1+\varepsilon_0\right)$$

$$M_{\infty,c} = M_{\infty,u}\left[1+\left(1+\frac{\kappa-1}{2}M_{\infty,u}^2\right)\varepsilon_0\right]$$

$$q_{\infty,c} = q_{\infty,u}\left[1+\left(2-M_{\infty,u}^2\right)\varepsilon_0\right]$$

$$p_{\infty,c} = p_{\infty,u}\left(1-\kappa M_{\infty,u}^2\varepsilon_0\right) \tag{7.40}$$

$$T_{\infty,c} = T_{\infty,u}\left(1-(\kappa-1)M_{\infty,u}^2\varepsilon_0\right)$$

$$\rho_{\infty,c} = \rho_{\infty,u}\left(1-M_{\infty,u}^2\varepsilon_0\right)$$

$$Re_{\infty,c} = Re_{\infty,u}\left[1+\left(1-\frac{\kappa}{2}M_{\infty,u}^2\right)\varepsilon_0\right]$$

The dimensionless aerodynamic coefficients are analogously obtained as

$$c_{p,c} = \frac{p-p_{\infty,c}}{\frac{1}{2}\kappa p_{\infty,c}M_{\infty,c}^2} = \frac{p-p_{\infty,u}\left(1-\kappa M_{\infty,u}^2\right)}{\frac{1}{2}\kappa p_{\infty,u}M_{\infty,u}^2\left(1-\kappa M_{\infty,u}^2\right)\left[1+\left(1+\frac{\kappa-1}{2}M_{\infty,u}^2\right)\varepsilon_0\right]^2}$$

$$C_{L,c} = \frac{L}{\frac{1}{2}\kappa p_{\infty,c}M_{\infty,c}^2 A_{ref}} = \frac{L}{\frac{1}{2}\kappa p_{\infty,u}M_{\infty,u}^2 A_{ref}}\frac{1}{\left(1-\kappa M_{\infty,u}^2\right)\left[1+\left(1+\frac{\kappa-1}{2}M_{\infty,u}^2\right)\varepsilon_0\right]^2}$$

$$C_{D,c} = \frac{D}{\frac{1}{2}\kappa p_{\infty,c}M_{\infty,c}^2 A_{ref}} = \frac{D}{\frac{1}{2}\kappa p_{\infty,u}M_{\infty,u}^2 A_{ref}}\frac{1}{\left(1-\kappa M_{\infty,u}^2\right)\left[1+\left(1+\frac{\kappa-1}{2}M_{\infty,u}^2\right)\varepsilon_0\right]^2}$$

$$C_{m,c} = \frac{M_y}{\frac{1}{2}\kappa p_{\infty,c}M_{\infty,c}^2 A_{ref}C_{ref}} = \frac{M_y}{\frac{1}{2}\kappa p_{\infty,u}M_{\infty,u}^2 A_{ref}C_{ref}}\frac{1}{\left(1-\kappa M_{\infty,u}^2\right)\left[1+\left(1+\frac{\kappa-1}{2}M_{\infty,u}^2\right)\varepsilon_0\right]^2}.$$

$$\tag{7.41}$$

### 7.1.3.3 Corrections for Wake Blockage

Due to viscous effects, the disturbance of the model does not decay along the wind tunnel circuit. As Figure 7.13 illustrates, the wake past the model due to the velocity defect leads to a blockage disturbance. For the analysis, the appropriate singularity is a single source with a strength scaling with the drag of the model, as it is closely linked to the displacement thickness of the wake shear layer

$$q = \frac{1}{2}U_\infty \cdot c \cdot C_D. \tag{7.42}$$

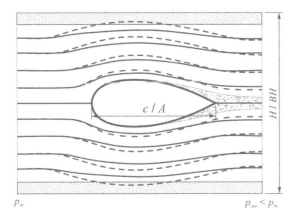

**FIGURE 7.13**  Effect of wake blockage on flow field in a closed test section.

For the source placed in the wind tunnel axis, the corresponding streamwise influence factor results in

$$\varepsilon_w = \frac{C_D}{4\pi\beta^2} \sum_{\substack{n=-\infty \\ n\neq 0}}^{n=\infty} (-1)^n \frac{\xi}{\xi^2 + (\zeta - n)^2}. \tag{7.43}$$

Krynytzky [11] explains that "downstream of the model, the tunnel cross-sectional area is decreased by the equivalent displacement area of the viscous wake plume so that the flow external to the wake must increase proportionately. In total, the image sources add additional mass to the oncoming stream, so that the uniform velocities far upstream and downstream cannot be equal. An interesting result for this singularity set is the non-zero interference far upstream of the model" – (Krynytzky in [11] p. 2-26) (see Figure 7.14).

It is important to note that the streamwise interference by wake blockage at the model position is zero. But, since the velocity of the tunnel is usually determined by pressure measurements far upstream, the streamwise interference of the wake blockage acts on these pressure readings. It is, therefore, appropriate to regard the interference at the model position as the correction towards the measured tunnel flow speed. The streamwise interference factor for two-dimensional airfoil models is then given by

$$\varepsilon_w(0,0) = \frac{C_D}{4\beta^2} \frac{c}{H} \tag{7.44}$$

and for three-dimensional models by

$$\varepsilon_w(0,0,0) = \frac{C_D A}{4\beta^2 BH} \tag{7.45}$$

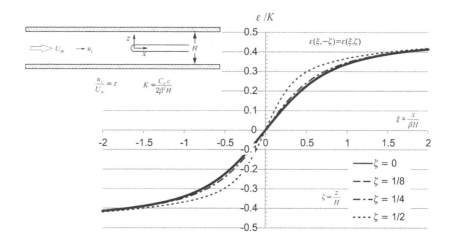

**FIGURE 7.14**   Streamwise interference of a 2D source in a closed-wall tunnel. (According to Krynytzky in [11], figure 2.18.)

$$\frac{\partial \varepsilon_w (0,0,0)}{\partial \xi} = \frac{\pi}{12} \frac{C_D}{\beta^2} \frac{c}{H}. \tag{7.46}$$

### 7.1.3.4   Model Size Considerations

From the discussion of wind tunnel corrections, it is obvious that the size of the model has a severe impact on the measured quantities. In order to keep the magnitude of wind tunnel corrections at an acceptable limit, the model size should not exceed certain limits. "It should be noted that aerodynamic size of the model, which depends on the dominant flow phenomena and on the magnitude of the generated aerodynamic forces (see next section), rather than geometric size, is the most relevant characterisation of model size" – (Krynytzky & Hackett in [11] p. 1-19). Mainly the representation of the full aerodynamic body by single sources and the distribution of aerodynamic effects in space shall be considered. "With regard to the magnitude of the disturbances due to the model, V (volume; for 2D flows, cross-sectional area A) and $[C_D \times A_{ref}]$ (model drag; for 2D flows, $[c_D \times c_{ref}]$) are taken to be the relevant linear scaling parameters representing the symmetric displacement of far-field streamlines, and $[C_L \times A_{ref}]$ (model lift; for 2D flows, $[c_L \times c_{ref}]$) for the asymmetric far-field perturbation due to the model" – (Krynytzky & Hackett in [11] p. 1-19).

"Model size, as relating to blockage interference, is often described or delimited by the so-called 'model blockage' parameter, or $A_{max}/C$, where $A_{max}$ is the maximum model cross-sectional area (taken normal to the tunnel axis), and $C$ is the test section cross-sectional area. The 2D equivalent is $t_{max}/H$, where $t_{max}$ is the maximum model thickness and $H$ is the test section height" – (Krynytzky & Hackett in [11] p. 1-19). According to Hackett [17] "it is shown that, up to an $[A_{ref}/C]$ value of approximately 10%, corrected drag coefficients lie within the scatter band of accepted constraint-free data" – (Hackett [17] p. 5).[7] For transport aircraft with an approximate airfoil thickness of about 10%, this results in a model blockage parameter of about $A_{max}/C = 1\%$.

The effect of lift is slightly less predictable as the lift variation is not only related to model size. Some guidelines can be found regarding the span of a model wing. "Models with large span-to-tunnel-width ratio (exceeding about 0.6) will be particularly affected by tunnel-wall upwash interference varying along the span ..." – (Steinle & Stanewsky [18] p. 12). Rogers states that "though precision is not possible in such matters a 'small' wing may conveniently be regarded as one in which the ratio of the wing span [$b$] to the tunnel breadth [$B$] is below about 0.5" – (Rogers [10] p. 298). According to Krynytzky, "the effect of span can be ignored for span ratios less than about 0.5" – (Krynytzky in [11] p. 2-20).

For two-dimensional high-lift models exhibiting an even higher lift potential, the lift induction is not span-depending. According to van den Berg [19], "the use of relatively large two-dimensional models is not recommended for high-lift tests. The model-chord/tunnel-height ratio should never be chosen larger than $[c_{ref}/H] = 0.30$. If large flap angles are applied, a model chord/tunnel height ratio $[c_{ref}/H] = 0.25$ is more advisable" – (van den Berg [19] p. 5-5).

### 7.1.4  NON-CORRECTABLE SOURCES OF MEASUREMENT ERRORS

The presence of wind tunnel walls results in the build-up of a boundary layer along the wall. In straight test sections, this boundary layer has a displacement effect and, therefore, reduces the effective cross-section of the tunnel. Since the development of the boundary layer is not only dependent on the flow velocity but also on the pressure field induced by the wind tunnel model, the displacement thickness of the wall boundary layer cannot be predicted a priori.

#### 7.1.4.1  Side-Wall Separation at Two-Dimensional Airfoils

For two-dimensional airfoil models, Garner [10] remarks "that the boundary layers on the side walls do not have extensive influence on the pressure distribution. [...] the loss of total load on an aerofoil spanning a closed rectangular tunnel is unlikely to exceed 1% at incidences below the stall. [...] very close to the wall [...] the local lift is less than 10% below that at the centre of the tunnel [...]. Large changes in side-wall boundary-layer thickness are found to produce only small changes in the loading. One may therefore have confidence in a purely two-dimensional theoretical analysis" – (Garner in [10] p. 31). Mokry et al. [2] report that the displacement of the boundary layer mainly results in a change of Mach number

$$M_c = \left[ M_{\infty,u}^2 - \left( 2 + \frac{1}{H} - M_{\infty,u}^2 \right) \frac{2\delta_1}{b} \right]^{1/2} \tag{7.47}$$

and the correction of the lift coefficient can be approximated by the Prandtl-Glauert rule

$$C_{L,c} = C_{L,u} \sqrt{\frac{1 - M_\infty^2}{1 - M_c^2}}. \tag{7.48}$$

Nevertheless, for flapped airfoil models and investigations of stall onset, the weakened boundary layer flow at the junction between wind tunnel walls and the model produces a premature separation of the corner flow. "Since the side wall boundary layer is much thicker than that on the model, it is more prone to separate when facing the same adverse pressure gradient generated at the rear portion of the airfoil. Separation cells formed at the airfoil-wall junction can be observed while the pressure gradient is not yet severe enough to cause separation of the boundary layer over the model" – (Mokry et al. [2] p. 159). "At larger geometrical aspect-ratios, there is a very considerable indirect influence of the flow separations near the walls on the measured forces at the mid-span section. This indirect influence is due to the large local lift losses that are associated with the flow separations. The local lift losses near the walls lead to a reduction of the effective aspect-ratio of the model, so that the geometrical incidence of the model in the tunnel will increase faster than the effective incidence at midspan. It is possible that the effective incidence even decreases with increasing geometrical incidence" – (van den Berg [19] p. 5-5).

Figure 7.15(a) shows the surface streak line pattern of a two-element high-lift airfoil model at high incidence. The side-wall separation has already started as clearly seen in the marked region which spans about 10% of the wing span. Figure 7.15(b) shows the corresponding lift curves for the same model derived from pressure integration in the center wing section. There is an obvious kink in the development of the lift coefficient showing a reduced lift slope. Additionally, the lift integration shows a higher uncertainty as the side-wall separation is not a fully steady flow. The onset of the corner separation induces a significant spanwise loading with a corresponding development of induced flow. As for the theory of finite span wings, this results in an induced flow angle at the center section that is less than the geometric incidence. Since the induced angle of attack is depending on the circulation, a reduced slope of the lift curve is observed.

The prevention of side-wall separation is possible by stabilizing the near wall boundary layer. For example, in Figure 7.15(b) a droop nose type device has been implemented at the sides of the wind tunnel model [20]. The lift curve shows a linear slope close to maximum lift coefficient. A comparison of the pressure distributions at the stall condition unveils that the flow characteristics are identical in both cases. This concludes that the flow in stalling condition at the center section is similar, which is not astonishing as it corresponds to a specific loading of the boundary layer. Anyhow, this is achieved at different angles of attack. As the pressure distribution is identical but the orientation of the airfoil is different, this leads to a deviation of the maximum lift coefficient value itself.

The side-wall separation, in principle, can also be prevented by the application of boundary layer control measures as discussed in Section 6.1. Merely, suction and tangential blowing on the side wall ahead of the model wall junction have been used in the past. Mokry et al. [2] prefer suction at the wind tunnel walls around or ahead of the junction of the airfoil model and the wall as "...the suction required [...] is moderate and does not remove the boundary layer completely. [...] With the boundary layer energized by suction, early separation at the side wall in the region of adverse pressure gradient can also be delayed. [...] Reducing the boundary layer thickness, however, does not imply full control of the boundary layer development [...]. It

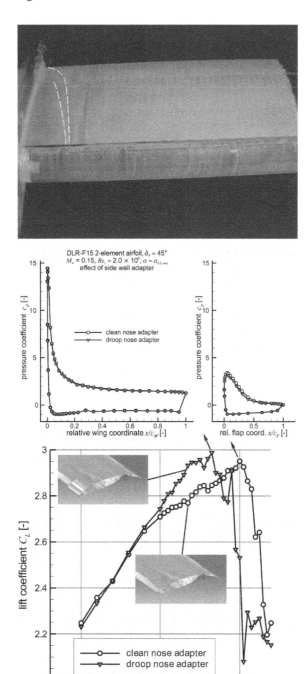

**FIGURE 7.15** Impact of side-wall separation on the measurements of highly loaded two-dimensional high-lift airfoil models. ((b) from Wild et al. [20], figure 7.)

should be noted that the boundary layer recovers to that of a flat plate rapidly, once it leaves the suction area and thus reacts to the model pressure field in a similar manner as that without suction" – (Mokry et al. [2] p. 161).

Examples of wall blowing are given by van den Berg [19] and de Vos [21]. In principle, they conclude blowing a more cost-efficient way and resume that a "fundamental advantage of boundary-layer control by blowing is that flow separations are prevented over a large area behind the slot, so that the position of the slot is not very critical" – (van den Berg [19] p. 5-8). Additionally, once a sufficient blowing is established, "the test results obtained at the mid-section of the model are virtually insensitive to increases in the blowing amounts" – (de Vos [21] p. 25). Still, a warning has to be given as the above statements are given towards lift coefficients and stall onset determination. The development of drag coefficients is strongly influenced by the amount of blowing, as data in [21] indicates.

### 7.1.4.2   Boundary Layer Influence on Half Model Testing

The benefit of using half models especially for the experimental investigation of high-lift systems has already been mentioned in Section 7.1.2. The usage of half-models bears the principal "difficulty to satisfy the physical boundary conditions in the sectional plane" – (Franz [9] p. 10). To be a perfect simulation, it is necessary to impose the symmetry of the flow in the middle plane of the aircraft fuselage. "Additional forces must not be applied to the model, neither by the space between weighed and the tunnel-fixed part of the arrangement nor by an elastic closing of this slot. The flow around the half-model should be as free as possible from the effect of the friction layer of the boundary wall. Additionally, it is necessary to decouple any flow forces on a mounting system from the balance measurements" – (Franz [9] p. 10).

Mounting the model directly on the wind tunnel wall imposes that the fuselage flow is dominated by the wind tunnel wall boundary layer. A possible way to eliminate this is to mount the model on a support and to implement a splitter plate, or "false wall," at the fuselage midsection. This arrangement bears the difficulty of eliminating the forces on the wall-pointing side from the aerodynamic force measurements. Additionally, there is a significant gradient of the boundary layer thickness along the fuselage.

Current state of the art is the application of a cylindrical offset shaped with the center cross-section of the fuselage, the so-called "peniche" (French: tow barge), which is mechanically decoupled from the force measurements. The arrangement is illustrated in Figure 7.16. "The fuselage sectional plane and thus also the internal limitation of the 'weighed' model parts are thus removed by this device from the friction-afflicted tunnel wall if the height of the pedestal corresponds approximately to the maximum boundary layer thickness at the fuselage rear" – (Franz [9] p. 12).

A consequence of the cylindrical offset is a change in the effective aspect ratio of the model by the virtual extension of the span. Melber and Wichmann [22] have investigated the influence of the peniche height on the measurements of half-models of aircraft with high-lift systems. Their findings conclude that "the main peniche effect on the model flow is based on its additional flow displacement leading to an additional flow velocity around the fuselage and the inboard wing compared to a

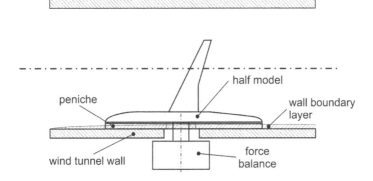

**FIGURE 7.16** Arrangement of half-model inside test section applying cylindrical offset (peniche) to reduce the impact of the wind tunnel wall boundary layer.

configuration without a peniche. The strength of the peniche displacement effect is directly linked to the angle of attack of the configuration by means of a lift rise with increasing peniche height growing with increasing angle of attack" – (Melber & Wichmann [22] p. 13).

### 7.1.4.3   Model Deformation

The aim of wind tunnel experiments on full aircraft configurations, either as half-model or as full-model, is the simulation of the flow as similar as possible compared to the real flight condition. In any case, the mechanical construction under significant load is not rigid and will exhibit elastic deformations.

The aerodynamic loads generated by pressure dominate in most cases. They scale with the dynamic pressure of the flow and the square of the model scale, as the force scales with the surface.

$$F \sim \left( p_\infty \cdot M_\infty^2, A_{ref} \right) \stackrel{\wedge}{=} scale^2. \tag{7.49}$$

Moments scale additionally with the model scale due to the lever arm

$$M \sim \left( p_\infty \cdot M_\infty^2, A_{ref} \cdot c_{ref} \right) \stackrel{\wedge}{=} scale^3. \tag{7.50}$$

The planar bending deformation of a beam is defined by the differential equation [23]

$$\frac{d^2x}{dy^2} + \frac{M_b(x)}{E \cdot I_{ii}(x)} = 0. \tag{7.51}$$

The maximum deformation scale for force-induced bending with the ratio of the force and the surface inertia, and the cube of the length, resulting in the proportionality

$$\delta \sim \frac{F \cdot l^3}{I_{ii}} \stackrel{\wedge}{=} \frac{scale^5}{scale^4} = scale. \tag{7.52}$$

The torsion deformation is defined by the differential equation

$$\frac{d\varphi}{dx} = \frac{M_t(x)}{G \cdot I_p(x)} \tag{7.53}$$

and scales with the ratio of the torsion moment and the polar surface inertia

$$\varphi \sim \frac{M \cdot l}{I_p} \triangleq \frac{scale^4}{scale^4} = 1. \tag{7.54}$$

To ideally reproduce the bending and torsion deformation, it is, therefore, necessary to truly scale the structural layout of the aircraft. This is impractical, as the shell thickness gets too thin for local deformations like buckling, connections by rivets cannot be properly scaled, and further manufacturing constraints. In general, a reproduction of the wing structure in model scale size will result in a stiffer wing in relation to the real aircraft.

Therefore, the model shape in the wind tunnel needs to take the designated deformation into account. In practice, wind tunnel models are built as stiff as possible to minimize the deformation in the wind tunnel test. Figure 7.17 shows a comparison of the principal structural layout of a real aircraft wing and a wind tunnel model wing. To achieve a mostly correct simulation of the real aircraft behavior, the wing shape includes the theoretical deformation of the aircraft wing under flight conditions. It is obvious that the model can only be built according to one flight condition deformation, so deviations from the real shape in another flight or load condition of the aircraft must be respected in the analysis and interpretation of measurement data.

In case measurements are made in a pressurized tunnel, the different deformation behavior of the model and real aircraft wings has to be considered. A proper analysis can only be done by measuring the model deformation during the tests. An optimal way to achieve a best correspondence between model and aircraft is to take into account the model deformation in the wind tunnel model design, namely to design the shape of the model so that, under wind load in the tunnel, the deformed model achieves the designated shape of the aircraft under real flight conditions, as, e.g., demonstrated by Bier et al. [24].

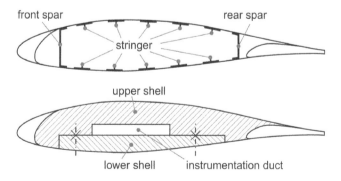

**FIGURE 7.17**   Comparison of real aircraft (upper) and wind tunnel model (lower) wing structure.

## 7.2 NUMERICAL SIMULATION

Numerical simulation in this context means any method that simulates the behavior of the aerodynamics by mathematical models. It is not the intention of this section to give a full description of the available mathematical and numerical modeling. It shall only give a rough overview with the emphasis to understand, which methods are suited for the simulation of high-lift flows. It is intended to discuss mainly the general idea of the methodologies. *The following discussion is therefore in no sense complete, neither mathematically nor in the different details of the methodologies.* Interested readers are referred to specific textbooks, e.g., [25].

Nevertheless, all simulation methods attempt to describe the aerodynamics by calculating the fluid properties around the aircraft, namely density, pressure, and velocities. The motion of air is governed by the Navier-Stokes equations for a Newtonian fluid. The equations describe the spatial and temporal variation of the fluid properties in a volume by formulating the corresponding conservation laws for mass, momentum, and energy.

$$\text{mass: } \frac{\partial}{\partial t}\oiiint_V \rho\, dV + \underbrace{\oiint_\Omega \rho(\mathbf{u}\cdot\mathbf{n})\, d\Omega}_{\text{mass flux through boundary}} = 0$$

$$\text{momentum: } \frac{\partial}{\partial t}\oiiint_V \rho\mathbf{u}\, dV + \underbrace{\oiint_\Omega \rho\mathbf{u}(\mathbf{u}\cdot\mathbf{n})\, d\Omega}_{\substack{\text{momentum flux} \\ \text{through boundary}}} = \underbrace{\oiint_\Omega (-p\mathbf{n}+\tau\cdot\mathbf{n})\, d\Omega}_{\text{forces on boundary}}$$

$$\text{energy: } \frac{\partial}{\partial t}\oiiint_V \rho E\, dV + \underbrace{\oiint_\Omega \rho H(\mathbf{u}\cdot\mathbf{n})\, d\Omega}_{\text{enthalpy flux through boundary}} = \underbrace{\oiint_\Omega (\lambda_T \nabla T\cdot\mathbf{n})\, d\Omega}_{\text{heat flux}} + \underbrace{\oiint_\Omega ((\mathbf{u}^T\cdot\tau)\cdot\mathbf{n})\, d\Omega}_{\substack{\text{viscous energy} \\ \text{dissipation}}}$$

$$(7.55)$$

### 7.2.1 COMPUTATIONAL FLUID DYNAMICS (CFD)

The approach to numerically calculate the full equations lead to the method of Direct Numerical Simulation (DNS). In this approach, the spatial and temporal variation has to be fully resolved in the full volume and therefore leads to a very high numerical effort, both in computational memory and time. The approach is capable of fully simulating the time-dependent turbulent flow down to the smallest structures of fluid motion.

To reduce the computational effort, a distinction can be made between large turbulent structures influencing the time-dependent evolvement of aerodynamic forces and small-scale structures that are mostly affecting the dissipation of energy. This kind of method is called Large Eddy Simulation (LES) and needs an additional model for the dissipation of turbulence into heat. The size of the resolved turbulent structures mainly depends on the density of the discretization.

For many relevant aerodynamic conditions at the aircraft, the perceived aerodynamic forces are steady in time, although the flow itself is not. An approach to achieve a steady-state simulation while respecting turbulent flow arises from the Reynolds-averaged Navier-Stokes (RANS) equations. In this description of the fluid dynamics, every fluid property is decomposed into a steady-state mean value and a timely variation with a zero-mean value

$$\phi(t) = \bar{\phi} + \phi'(t); \quad \overline{\phi'(t)} = 0. \tag{7.56}$$

The derived set of equations is similar to the Navier-Stokes equations for the mean values with the extension of a second-order closure term, the so-called Reynolds-stress, to respect the average forces and energy dissipation created by turbulence. The corresponding RANS methods, therefore, need a turbulence model to simulate the time-averaged impact of turbulent fluctuations. In comparison to LES methods, the effort is largely reduced as the spatial resolution is no longer driven by the size of turbulent structures. Additionally, there is no direct need for a certain temporal resolution as these methods only calculate the steady mean state.

It is possible to keep in the RANS approach the temporal variation of the mean value of the fluid properties. In this way, the unsteady RANS (URANS) approach is able to simulate the unsteady evolution of aerodynamic flows. In contrast to DNS and LES, the timely resolution does not need to resolve the turbulent structures but only the variation of the mean flow. This method is especially suitable for motions that are much slower than the turbulent motion itself, e.g., pitching airfoils, rotor flows, or aircraft maneuvers.

A beneficial combination of methods is the mixing of LES methods and RANS methods, called Detached Eddy Simulation (DES). In this methodology, the wall boundary layer with its high frequency small-scale turbulent structures is simulated by RANS to take advantage of the lower requirements on the spatial resolution. In the outer flow field, an LES approach allows for the coverage of the very large turbulent structures. These methods are specially designated to simulate highly separated flows and flows around bluff bodies.

A simplification that allows drastically reducing the computational effort is to neglect viscous effects. Removing the friction contribution from the Navier-Stokes equations leads to the Euler equations

$$\text{mass: } \frac{\partial}{\partial t} \iiint_V \rho \, dV + \underbrace{\oiint_\Omega \rho (\mathbf{u} \cdot \mathbf{n}) d\Omega}_{\text{mass flux through boundary}} = 0$$

$$\text{momentum: } \frac{\partial}{\partial t} \iiint_V \rho \mathbf{u} \, dV + \underbrace{\oiint_\Omega \rho \mathbf{u} (\mathbf{u} \cdot \mathbf{n}) d\Omega}_{\substack{\text{momentum flux} \\ \text{through boundary}}} = \underbrace{\oiint_\Omega -p\mathbf{n} \, d\Omega}_{\text{forces on boundary}} \tag{7.57}$$

$$\text{energy: } \frac{\partial}{\partial t} \iiint_V \rho E \, dV + \underbrace{\oiint_\Omega \rho H (\mathbf{u} \cdot \mathbf{n}) d\Omega}_{\text{enthalpy flux through boundary}} = \underbrace{\oiint_\Omega (\lambda_T \nabla T \cdot \mathbf{n}) d\Omega}_{\text{heat flux}}$$

and the corresponding methods are called Euler-methods. These methods give reasonable results for flows where pressure forces and compressibility effects dominate the flow behavior, which is mainly the case for flows in the transonic regime.

Inherent to all modeling approaches is that the continuous variation of flow properties is discretized into more or less small pieces where the property within these elements is assumed to be constant or to vary linearly. This approach allows transforming differentials and integrals into differences and sums, respectively.

$$\int \phi\, dx \rightarrow \sum_i \phi_i \Delta x_i$$

$$\frac{\partial \phi}{\partial x} \rightarrow \frac{\phi_{i+1} - \phi_{i-1}}{2\Delta x_i} (\text{central difference around } i)$$

(7.58)

Two principal approaches are used to achieve a spatial discretization, called mesh or grid, of the flow volume. Figure 7.18 outlines the principal arrangement of the two common methods of spatial discretization. The so-called structured grid approach divides the volume into boundary conforming quadrilateral (in 3D hexahedral) cells. The structuring achieves a regular subdivision with a known connectivity and

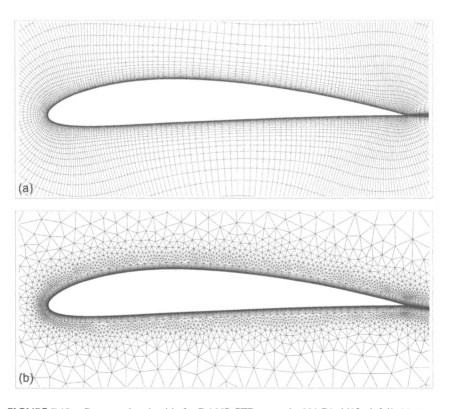

**FIGURE 7.18** Computational grids for RANS CFD around a NACA 4412 airfoil: (a) structured, (b) unstructured.

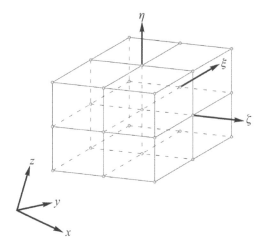

**FIGURE 7.19** Local coordinates in the computational space of a structured mesh discretization.

a constant defined number of neighboring cells. The so-called unstructured grid approach uses triangular cells (tetrahedrons in 3D). The unstructured approach is more flexible to generate grids around arbitrary configurations, while the structured grid approach offers especially a high accuracy for the resolution of boundary layers due to the achievable very high stretching of the grid cells. To benefit from the advantages of both methodologies, hybrid methods are commonly in use that use body-fitted quasi-structured layers close to the geometry to resolve the boundary layers and unstructured cells on the outer flow field.

Explained in Figure 7.19 by the structured grid, the approach for flow simulation by CFD transforms the physical space into the computational space by using local coordinates along the grid lines. This pre-processing allows that all metric terms that define the relation between physical and computational space are evaluated only once for a given grid a priori to the computations of the flow for various flow conditions.

For every cell, the discretized equations have to be solved in the transformed coordinate system. This leads to a large system of linear equations, which is solved iteratively by computing the fluxes between the different cells and evaluate the conservation laws in the chosen modeling approach.

RANS-based methods that were introduced in the mid-1990s are the major methods to analyze the aerodynamics today. "[...] surface pressures, skin friction, lift, and drag can generally be predicted with reasonably good accuracy at angles of attack below stall" – (Rumsey & Ying [26], p. 173). In recent years, some comparative efforts have been taken to qualify RANS-based CFD for the prediction of high-lift flows including stall, known as the AIAA High-Lift Prediction Workshops (HiLiPW). A major conclusion was that "in general, CFD [tends] to underpredict lift, drag, and the magnitude of the moment [...] compared with experiment. Predicting the flow [is] more difficult at higher angles of attack nearing stall" – (Rumsey et al. [27], p. 2078). Especially, the prediction of the onset of local separations triggering stall is sensitive to the chosen turbulence modeling, the spatial discretization type, and resolution. In this context, it is conclusive that "the inconsistencies between the

CFD results tended to be larger on the flap as well as at the outboard stations of the wing" – (Rumsey & Slotnick [28] p. 1023). Even past the third workshop, it has been stated that "accurate computations near maximum lift conditions remain collectively elusive" – (Rumsey et al. [29] p. 642).

On the other hand, methods like DES, LES, and DNS are expected to improve the prediction as the uncertainty imposed by turbulence modeling is reduced or even avoided. Nevertheless, the computational effort for these methods is still too high for available computational resources in a routine way, so only exemplary applications, e.g., by Escobar et al. [30], have been achieved up to now.

### 7.2.2 PANEL METHOD

Further simplifying the Euler equations eq. (7.57) by assuming incompressible and irrotational flow establishes the linearized potential equation

$$\Phi_{xx} + \Phi_{yy} + \Phi_{zz} = 0. \tag{7.59}$$

The method to solve this equation in a discretized way for arbitrary bodies has been introduced by Hess & Smith [31]. In contrast to CFD methods described above, it is sufficient to solve the system of equations only at the surface of the body. Instead of the need to discretize the full volume, only the body surface needs to be divided into pieces, the so-called panels from which the name of the method is derived: the panel method. A comprehensive overview of the panel method, its variants, and its implementation are given by Katz & Plotkin [32]. In the following, only the simplest variant is described to get an insight into the way of working.

Figure 7.20 depicts a very simple example of a two-dimensional panel method, where the discrete panels describing the airfoil are distributed source singularities of constant emissivity. The disturbance potential by a panel at an arbitrary point in the flow field is given by

$$\Phi_i = \frac{q_i}{2\pi} \int_0^1 \ln\left(r(s)^2\right) ds$$

$$r(s) = \sqrt{x_i(s)^2 + z_i(s)^2} \tag{7.60}$$

$$\begin{pmatrix} x_i(s) \\ z_i(s) \end{pmatrix} = \begin{pmatrix} x_i \\ z_i \end{pmatrix} + s \begin{pmatrix} x_{i+1} - x_i \\ z_{i+1} - z_i \end{pmatrix}.$$

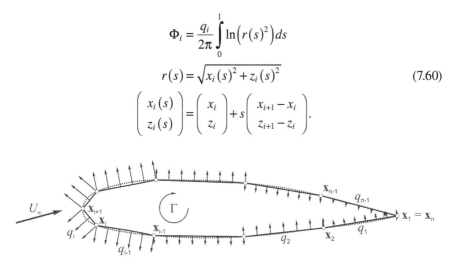

**FIGURE 7.20** Sketch of the principal panelization of an airfoil in a simple panel method.

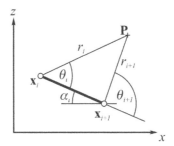

**FIGURE 7.21**   Nomenclature for the derivation of induced velocities by a panel source.

The induced velocity at an arbitrary point is then given by

$$u_i(x,z) = \frac{q_i}{2\pi} \int_0^1 \frac{x - x_i(s)}{\left[x - x_i(s)\right]^2 + \left[z - z_i(s)\right]^2} ds$$

$$w_i(x,z) = \frac{q_i}{2\pi} \int_0^1 \frac{z - z_i(s)}{\left[x - x_i(s)\right]^2 + \left[z - z_i(s)\right]^2} ds,$$

(7.61)

which are according to [32] and the definitions depicted in Figure 7.21.

$$\mathbf{u}_i = \begin{pmatrix} \cos\alpha_i & \sin\alpha_i \\ -\sin\alpha_i & \cos\alpha_i \end{pmatrix} \frac{q_i}{2\pi} \begin{pmatrix} \ln\left(\dfrac{r_i}{r_{i+1}}\right) \\ (\theta_{i+1} - \theta_i) \end{pmatrix}.$$

(7.62)

The integrals only depend on the geometry and, like for CFD, can be pre-processed ahead of any following flow simulation for different flow conditions.

The flow potential of the lifting vortex located at quarter chord is defined by

$$\Phi_\Gamma = \frac{\Gamma}{2\pi} \tan^{-1}\left(\frac{z}{x - 0.25}\right)$$

(7.63)

and the induced velocity at an arbitrary point is given by

$$u_\Gamma(x,z) = -\frac{\Gamma}{2\pi} \frac{z}{(x - 0.25)^2 + z^2}$$

$$w_\Gamma(x,z) = -\frac{\Gamma}{2\pi} \frac{(x - 0.25)}{(x - 0.25)^2 + z^2}.$$

(7.64)

In order to solve for the source distribution and the circulation, a set of equations is defined prescribing the flow conditions at the boundaries. First, a tangential flow condition is imposed at the center of each of the n-1 panels

$$\left[ \mathbf{U}_\infty + \mathbf{u}_\Gamma \left( \mathbf{x}_{i+\frac{1}{2}} \right) + \sum_{i=1}^{n-1} \mathbf{u}_i \left( \mathbf{x}_{i+\frac{1}{2}} \right) \right] \cdot \mathbf{n}_i = \mathbf{0}. \tag{7.65}$$

Second, the Kutta-condition prescribing the rear stagnation point at the trailing edge

$$\mathbf{U}_\infty + \mathbf{u}_\Gamma \left( \mathbf{x}_1 = \mathbf{x}_n \right) + \sum_{i=1}^{n-1} \mathbf{u}_i \left( \mathbf{x}_1 = \mathbf{x}_n \right) = 0 \tag{7.66}$$

provides the missing nth equation to define a closed linear system of equations. Improvements to panel methods can be obtained by using higher order continuous distributions of the source strength along with the panels, distributed vorticity at the panels, or application of distributed dipoles instead of vortices.

As discussed in Section 5.2, the major effect of lift augmentation by multi-element airfoils can be explained based on potential flow theory. It is, therefore, not astonishing that panel methods are quite appropriate to simulate the flow around high-lift systems as long as the flow is not dominated by the viscous effects, namely, flow separation. It is possible to predict pressure distributions and lift forces in the linear range of the lift curve quite well.

The distinct drawback of panel methods for the prediction of airfoil properties with regard to drag is the assumption of inviscid flow inherent to potential theory. Additionally, for high-lift flows, the onset of separation limiting lift properties is not directly possible. An approach to partially overcome this deficit is a coupling to one-dimensional boundary layer computations (see [33, 34]). These inviscid-viscous coupled methods, called Panel-Boundary-Layer methods, iteratively solve for the flow considering the displacement and momentum loss of the flow by the viscous shear forces. Based on a first inviscid computation, these methods solve the one-dimensional boundary layer momentum equation either in differential (eq. (2.49)) or integral (eq. (2.60)) form for a given velocity distribution. The coupling to the inviscid panel method can be obtained in two ways. Either the geometry nodes are displaced by the momentum loss thickness so that the geometry nodes $\mathbf{x}_i$ are moved according to Figure 7.22 to

$$\mathbf{x}'_i = \mathbf{x}_i + \delta_2 \left( \mathbf{x}_i \right) \cdot \frac{1}{2} \left( \mathbf{n}_{i-1} + \mathbf{n}_i \right) \tag{7.67}$$

or the tangential flow condition according to eq. (7.65) is imposed at a virtual point displaced by the boundary layer momentum loss thickness

$$\mathbf{x}'_{i+\frac{1}{2}} = \mathbf{x}_{i+\frac{1}{2}} + \delta_2 \cdot \mathbf{n}_i \tag{7.68}$$

obtaining a transpiration velocity directly at the panel, as illustrated in Figure 7.23.

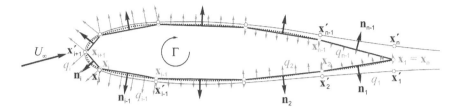

**FIGURE 7.22**  Principle of panel-boundary layer method based on moved nodes.

The ability to predict maximum lift coefficients or separation onset behavior strongly depends on the reliability of the chosen method to calculate boundary layer. Some implementations have been successfully reported by Cebeci et al. [35] and Dargel & Jacob [36]. Implementing separation onset criteria like those of Stratford discussed in Section 2.3 can further improve the simulation and separation prediction accuracy.

### 7.2.3 LIFTING-LINE METHOD

Prandtl introduced the lifting-line theory as a method to evaluate the spanwise variation of the lift distribution and the induced drag of a planar straight wing [37]. The lifting surface is represented by elementary horse-shoe vortices. Figure 7.24 sketches the vortex arrangement for a swept and tapered wing. According to Helmhotz's vortex rules, the circulation along these elementary vortices is constant. The relation between the circulation and the local sectional lift coefficient is given by Zhukovsky's theorem

$$\Gamma_i = \frac{1}{2} c_L(\alpha) \cdot U_\infty \cdot c_i. \tag{7.69}$$

Every elementary vortex induces a change in the effective incidence of the incoming flow. According to Prandtl, the induced angle of attack can be calculated by applying the Biot-Savart law by

$$\alpha_i(y) = -\frac{1}{4\pi U_\infty} \int_{-b/2}^{b/2} \frac{\partial \Gamma}{\partial y} \frac{1}{y - y'} dy'. \tag{7.70}$$

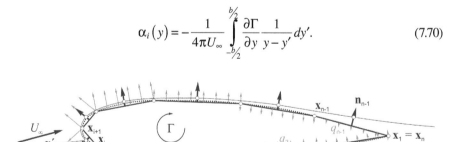

**FIGURE 7.23**  Principle of panel-boundary layer method based on transpiration velocity.

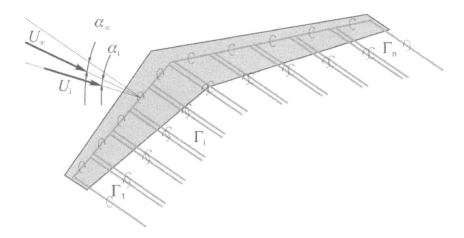

**FIGURE 7.24** Sketch of a lifting-line method establishing Prandtl's lifting line theory in discrete manner.

In the lifting-line method, the change of circulation is limited to the spanwise ends of the elementary vortices. The integral in eq. (7.70) is replaced by the sum equation

$$\alpha_i(y_i) = -\frac{1}{4\pi U_\infty} \sum_{\substack{j=1 \\ j \neq i}}^{n-1} \frac{\Gamma_{j+1} - \Gamma_j}{y_{j+1} - y_j} \frac{1}{y_i - y_j}. \tag{7.71}$$

The local incidence of the wing section is then given by

$$\alpha_{eff}(y_i) = \alpha_\infty + \alpha_i(y_i) + \varepsilon(y_i) \tag{7.72}$$

including the twist of the wing by the local twist angle $\varepsilon$ of the wing section.

The crucial element of the lifting-line method is the provision of the lift characteristics $c_L(\alpha)$ of the wing section. In the lifting-line method, a closed-loop starts with the induced incidences to zero, provides the sectional lift coefficient from the lift curve, calculates the circulation, and updates the induced incidence.

The method is independent of the source of the lift curve. It can be from experimental or any other numerical method. Jacob [38] has shown the implementation with 2D panel-boundary layer methods described above. Such a method has successfully been applied for high-lift design by Reckzeh [39]. Especially the coupling to two-dimensional lift characteristics obtained from experiments or by RANS methods allows for a reliable prediction of the lift distribution. In this case, the lift curve includes a reliable stall prediction per wing section. As depicted in Figure 7.25, by comparing the actual sectional lift coefficient with the maximum sectional lift coefficient of the corresponding airfoil, prediction of the spanwise stall onset is possible, as long as it is not triggered by three-dimensional effects like crossflow or disturbances not included in the lifting-line method, e.g., side edge vortices.

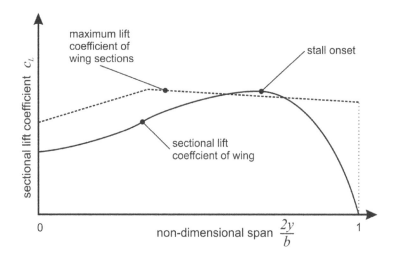

**FIGURE 7.25**   Lift distribution from lifting-line method indicating stall onset.

## NOTES

1. Named after the French inventor Gustav Eiffel, who first built a wind tunnel of this type at the foot of his Eiffel tower in Paris in 1909. In 1912, the tunnel was moved to Auteuil. The tunnel is again in service and mainly serves for investigations of building aerodynamics [40].
2. Named after the town where Ludwig Prandtl built up the first aerodynamic research facility in Germany in 1908.
3. Krynytzky in [11] eq. (2.13); the formula given by Krynytzky is erroneous as it includes the factor $1/4\pi$ instead of $1/2\pi$, although the diagrams are correct.
4. Krynytzky in [11] eq. (2.15); the formula given by Krynytzky is erroneous, as it includes the factor $1/4\pi$ instead of $1/2\pi$, although the diagrams are correct; additionally the notation of eq. (2.15) is inconsistent with the definition of the axial flow induction coefficient by eq. (2.6) by the denominator $H/(c \cdot C_L)$.
5. Krynytzky in [11] neglects the terms based on body shape factor and drag coefficient that are originally contained in the formulation of Allen & Vincenti [13]. In practical applications, this is sometimes advantageous as the accuracy of the body shape factor and an accurate measured drag coefficient might not be available.
6. Krynytzky in [11] eq. (2.46); the formula given by Krynytzky is erroneous as it misses the negative sign, although the diagram is correct.
7. it is often overlooked that Hackett [16] refers to the reference area $A_{ref}$ instead of the cross sectional area $A_{max}$.

## REFERENCES

[1] Glauert H (1933) Wind Tunnel Interference on Wings, Bodies and Airscrews, ARC R&M 1566.
[2] Mokry M, Chan Y Y, Jones D J (1983) Two-Dimensional Wind Tunnel Wall Interference, AGARDograph No. 281.
[3] ONERA (2018) F1 Low Speed Pressurized Wind Tunnel, ONERA Centre du Fauga-Mauzac, www.onera.fr/en/windtunnel.

[4] Lin JC, Dominik CJ (1997) Parametric Investigation of a High-Lift Airfoil at High Reynolds Numbers, Journal of Aircraft **34**(4), pp. 485–491.

[5] DNW (2018) The Cryogenic Wind Tunnel Cologne (KKK), Brochure German-Dutch Wind Tunnels, www.dnw.aero/media-center/downloads/brochures//download/4.

[6] Quest J (2000) ETW – High Quality Test Performance in Cryogenic Environment, 21st AIAA Aerodynamic Measurement Technology and Ground Testing Conference, AIAA paper 2000-2206.

[7] NASA (2015) National Transonic Facility Fact Sheet, http://www.nasa.gov/sites/default/files/atoms/files/m187007_ntfprint_508.pdf.

[8] Germain E, Quest J (2005) The Development and Application of Optical Measurement Techniques for High Reynolds Number Testing in Cryogenic Environment, 43rd AIAA Aerospace Sciences Meeting, AIAA Paper 2005-0458.

[9] Franz HP (1981) Die Halbmodelltechnik im Windkanal und ihre Anwendung bei der Entwicklung der Airbus Familie, Deutsche Gesellschaft fuer Luft – und Raumfahrt, Jahrestagung, Aachen, W. Germany, May 1114, 1981. 38 p. DGLR Paper 81-118, translation to English (1982) The Half-Model Technique in the Wind Tunnel and Its Employment in the Development of the Airbus Family, NASA-TM-76970.

[10] Garner HC, Rogers EWE, Acum WEA, Maskell EC (1966) Subsonic Wind Tunnel Wall Corrections, AGARD AG 109.

[11] Ewald BFR (Ed) (1998) Wind Tunnel Wall Correction, AGARD AG 336.

[12] Theodorsen T (19) The Theory of Wind Tunnel Wall Interference, NACA Report No. 410.

[13] Allen HJ, Vincenti WG (1944) Wall Interference in a Two-Dimensional-Flow Wind Tunnel, with Consideration of the Effect of Compressibility, NACA TR 782.

[14] Goldstein S (1942) Two-Dimensional Wind-Tunnel Interference. Part II, ARC R&M 1902.

[15] Havelock TH (1938) The Lift and Moment on a Flat Plate in a Stream of Finite Width, Proc. Roy. Soc., Series A, **166**(925), pp. 178–196.

[16] Batchelor GK (1944) Interference of Wings, Bodies and Airscrews in a Closed Tunnel of Octagonal Section, Report ACA-5.

[17] Hackett JE (1982) Living with Wind Tunnel Walls, 12th Aerodynamic Testing Conference, AIAA Paper 82-0583

[18] Steinle F, Stanewsky E (1982) Wind Tunnel Flow Quality and Data Accuracy Requirements, AGARD AR 184.

[19] van den Berg B (1971) Some Notes on Two-Dimensional High-Lift Tests in Wind Tunnels, In: Colin PE, Willams J, Assessment of Lift Augmentation Devices, AGARD LS 43-71.

[20] Wild J, Wichmann G, Haucke F, Peltzer I, Scholz P (2009) Large Scale Separation Flow Control Experiments within the German Flow Control Network, 47th AIAA Aerospace Sciences Meeting including The New Horizons Forum and Aerospace Exposition, AIAA Paper 2009-0530.

[21] de Vos DM (1973) Low Speed Windtunnel Measurements on a Twodimensional Flapped Wing Model using Tunnel Wall Boundary Layer Control at the Wing-Wall Junctions, NLR TR-70050-U.

[22] Melber-Wilkending S, Wichmann G (2009) Application of Advanced CFD Tools for High Reynolds Number Testing, 47th AIAA Aerospace Sciences Meeting including The New Horizons Forum and Aerospace Exposition, AIAA Paper 2009-0418.

[23] Ettemeyer A, Wallrapp O, Schäfer B (2006) Technische Mechanik, Manuscript for lecture at FH Munich.

[24] Bier N (2013) Design of a Wind Tunnel Model for Maximum Lift Predictions Based on Flight Test Data. 31st AIAA Applied Aerodynamics Conference, AIAA Paper 2013-2930.

[25] Anderson JD (1995) Computational Fluid Dynamics, McGraw-Hill, New York.
[26] Rumsey CL, Ying SX (2002) Prediction of High Lift: Review of Present CFD Capability, Progress in Aerospace Sciences **38**(2), pp. 145–180.
[27] Rumsey CL, Slotnick JP, Long M, Stuever RA, Wayman TR (2011) Summary of the First AIAA CFD High-Lift Prediction Workshop, Journal of Aircraft **48**(6), pp. 2068–2079, DOI: 10.2514/1.C031447.
[28] Rumsey CL, Slotnick JP (2015) Overview and Summary of the Second AIAA High-Lift Prediction Workshop, Journal of Aircraft **52**(4), pp. 1006–1025.
[29] Rumsey CL, Slotnick JP, Scalfani (2019) Overview and Summary of the Third AIAA High Lift Prediction Workshop, Journal of Aircraft **52**(4), pp. 621–644
[30] Escobar JA, Suarez CA, Silva C, López OD, Velandia JS, Lara CA (2015) Detached-Eddy Simulation of a Wide-Body Commercial Aircraft in High-Lift Configuration, Journal of Aircraft **52**(4), pp. 1112–1121.
[31] Hess JL, Smith AMO (1967) Calculation of Potential Flow about Arbitrary Bodies, Progress in Aeronautical Sciences **8**, Pergamon Press, pp. 1–138.
[32] Katz J, Plotkin A (2001) Low-Speed Aerodynamics, 2nd edition, Cambridge University Press, New York.
[33] Williams BR (1991) Viscous-Inviscid Interaction Schemes for External Aerodynamics. Sadhana **16**, pp. 101–140.
[34] Jacob K, Steinbach D (1974) A Method for Prediction of Lift for Multi-Element Airfoil Systems with Separation. AGARD CP-143.
[35] Cebeci T, Besnard E, Chen HH (1998) An Interactive Boundary-Layer Method for Multielement Airfoils, Computers & Fluids **27**(5–6), pp. 651–661.
[38] Dargel G, Jakob H (1988) Berechnung von Klappenprofilströmungen mit Ablösung auf der Basis gekoppelter Potential und Grenzschichtlösungen. Strömungen mit Ablösung. DGLR-Bericht 88-05, pp. 267–278.
[37] Schlichting H, Truckenbrodt E (2001), Aerodynamik des Flugzeuges, Band 1. Springer Verlag, Berlin, Heidelberg, New York.
[38] Jacob K (1993) A Fast Computing Method for the Flow Over High-Lift Wing, no. 23 in AGARD CP-515.
[39] Aérodynamique Eiffel (2021) Home page, https://www.aerodynamiqueeiffel.fr/ (accessed Oct 2021).
[40] Reckzeh D (1999) CFD-Methods for the Design Process of High-Lift Configurations, New Results on Numerical and Experimental Fluid Mechanics **72**, pp 347–354.

# 8 Aerodynamic Design of High-Lift Systems

## NOMENCLATURE

| | | | | | | |
|---|---|---|---|---|---|---|
| $b$ | $m$ | Span | $Re$ | – | Reynolds number |
| $c$ | $m$ | Chord length | $\mathbf{S}$ | – | Sweep transformation matrix |
| $c_{Df}$ | – | Section friction drag coefficient | $\mathbf{T}$ | – | Translation matrix |
| $c_{Dp}$ | – | Section pressure drag coefficient | $U$ | $m/s$ | Flow velocity |
| $c_L$ | – | Section lift coefficient | $V_S$ | $m/s$ | Stall speed |
| $c_p$ | – | Pressure coefficient | $\mathbf{x}$ | $m$ | Vector of design variables |
| $c_{ref}$ | – | Reference chord length | $x, y, z$ | $m$ | Cartesian coordinates |
| $C_D$ | – | Drag coefficient | $\tilde{x}, \tilde{y}$ | $m$ | Normalized coordinates |
| $C_{Df}$ | – | Friction drag coefficient | $x_c, y_c$ | $m$ | Conic center |
| $C_{Dp}$ | – | Pressure drag coefficient | $x_F, y_F$ | $m$ | Flap position |
| $C_L$ | – | Lift coefficient | $x_S, y_S$ | $m$ | Slat position |
| $C_{L,\max}$ | – | Maximum lift coefficient | $\alpha$ | $^\circ$ | Angle of attack |
| $D$ | $N$ | Drag force | $\gamma_{C_D}$ | – | Drag correlation coefficient |
| $f$ | – | Functional | $\gamma_{C_D}$ | – | Drag correlation coefficient |
| $f$ | – | Pressure correction factor (Lock [15]) | $\gamma_{TE}$ | $^\circ$ | Trailing edge opening angle |
| $F_{obj}$ | – | Objective function | $\Gamma$ | $m^2/s$ | Circulation |
| $g$ | – | Inequality constraint | $\delta$ | $^\circ$ | Deflection angle |
| $g_F$ | $m$ | Flap gap | $\delta_F$ | $^\circ$ | Flap deflection angle |
| $g_S$ | $m$ | Slat gap | $\delta_S$ | $^\circ$ | Slat deflection angle |
| $h$ | – | Equality constraint | $\eta$ | – | Dimensionless span $2y/b$ |
| Hi | – | Inverse Hessian | $\Lambda$ | – | Wing aspect ratio |
| $k$ | – | Weighting factor | $\varphi$ | $^\circ$ | Sweep angle |
| $s_{blend}$ | $m$ | Transition length | $\varphi_{eff}$ | $^\circ$ | Effective sweep angle |
| $L$ | $N$ | Lift force | $_{LE}$ | | Leading edge |
| $L/D$ | – | Lift to drag ratio | $_{opt}$ | | Optimal |
| $M$ | – | Mach number | $_{TE}$ | | Trailing edge |
| $ovl_F$ | $m$ | Flap overlap | $_\infty$ | | Onflow condition |
| $ovl_S$ | $m$ | Slat overlap | $_{2D}$ | | Two-dimensional |
| $\mathbf{R}$ | – | Rotation matrix | $_{2.5D}$ | | Swept wing transformation |
| $r$ | $m$ | Radius | $_{3D}$ | | Three-dimensional |

After having discussed the physical principles and limitations, the requirements and objectives, the principal solutions, and the methods for evaluation at the pre-product state, the design of high-lift systems poses the last and ultimate challenge.

DOI: 10.1201/9781003220459-8

It is important to understand that the design of high-lift systems is a very multi-disciplinary task. Rudolph [1] states, "sophisticated high-lift systems are heavy, expensive to build, and difficult to maintain" – (Rudolph [1] p. 137). As the high-lift system is only used during a very limited portion of flight, its impact on the overall characteristic is quite astonishing. For example, for a large twin-jet aircraft, benefits have been reported [2] to be in the order of:

1. A 0.10 increase in lift coefficient at constant angle of attack is equivalent to reducing the approach attitude by about one degree. For a given aft body-to-ground clearance angle, the landing gear may be shortened resulting in a weight saving of 1400 Ib [635 kg].
2. A 1.5% increase in maximum lift coefficient is equivalent to a 6600 lb [3000 kg] increase in payload at a fixed approach speed.
3. A 1% increase in take-off L/D is equivalent to a 2800 Ib [1270 kg] increase in payload or a 150 NM [278 km] increase in range.

**(Meredith [2] p. 19-1)**

As the list indicates, savings are primarily to be achieved via the increased lift capability and corresponding increase in payload or fuel. The weight of the high-lift system itself has therefore to be counterbalanced.

## 8.1  THE DESIGN PROCESS OF HIGH-LIFT SYSTEMS

Before looking into the details of a high-lift design, it is necessary to understand the design process of the whole aircraft. This design process is structured into three phases according to an ideal design process shown in Figure 8.1. The first step is a preliminary design of the aircraft concept. In this phase, the type of aircraft is specified in its global characteristics – e.g., number of passengers and mission – and its architecture – e.g., fuselage, wing, and propulsion type. Based on simplified analysis, basically handbook methods, the flight performance corresponding and the foreseen costs of the aircraft in terms of investment (non-recurring costs – NRC), operation (direct operating costs – DOC), and maintenance (recurring costs – RC) are estimated and detailed in the so-called Top-Level Aircraft Requirements

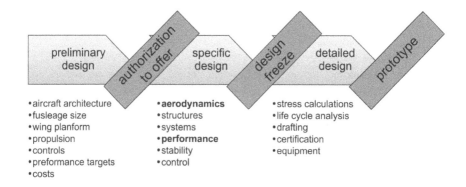

**FIGURE 8.1**  Aircraft design phases.

(TLARs). As aircraft development is a very costly undertaking it is important at the very beginning to find investors to finance the costs. Therefore, at the end of the preliminary design phase, the aircraft is sufficiently detailed to be offered to customers. For this reason, the end of the preliminary design phase is characterized by the Authorization to Offer (ATO), namely the decision of the manufacturer to sell the aircraft with the given characteristics to launch customers that contribute to the financing of the aircraft development by guaranteed orders. In conclusion, the TLARs are "the Bible" for the further design. Failing to achieve the estimated aircraft performance means to fall short behind a guaranteed economic value with corresponding additional costs. Achieving the TLARs is, therefore, the "live or die" of an aircraft project.

The second design phase – the specific design – now picks up the TLAR and is aimed to fulfill the requirements. The design is now going into the specific disciplines – majorly aerodynamics, structures, and systems. The single disciplines now optimize their degrees of freedom. While in former times, the disciplines worked more or less on their own, the interrelation was covered by multiple design loops exchanging design parameters, the trend is now on introducing a multidisciplinary understanding of the design problem. Understanding and respecting the targets of other disciplines at an early design stage leads to a speed-up of the design by reducing the number of design loops needed to achieve a converged result. When this converged result is achieved, and it fulfills the TLAR, the time has come to freeze the design and to start detailing.

The detailed design handles the transformation from an engineered design concept into a product. The most important parts surely are the stress calculations, the drafting for production, and all certification work that has to be completed before first flight. When this is completed, the manufacturing of the prototype of the aircraft is due.

Looking more into the details of the specific design of the wing, illustrated in Figure 8.2, the first step based on the TLARs and the wing planform is the design of the cruise shape. The definition of the full wing shape is most important not only to achieve the aircraft performance in cruise but also to define the available space for the definition of the internal structure and the integration of the high-lift system and other systems to be hosted inside the wing. Nevertheless, the achievement of the TLAR in terms of range and fuel consumption is the major objective in the design of the wing, which consists of the design of the base airfoils and the definition of the wing loft including thickness and twist distribution to achieve an optimal lift distribution for minimum induced and wave drag.

With the wing shape defined in a first step, the more specific designs are performed in parallel. The structural design focuses on the definition of the wing substructure, mainly the spar positions, the ribs and stringer spacing, and the wing shell thickness. The systems design takes care of the integration of all subsystems that are placed into the wing, including actuators for control systems, lights, and status monitoring including routing of all necessary wires and hydraulic lines. Especially on the inboard wing, the integration of the main landing gear is challenging and often in conflict with the wing shape. Finally, the high-lift devices have to be defined within the limited space of the given wing shell. At the end,

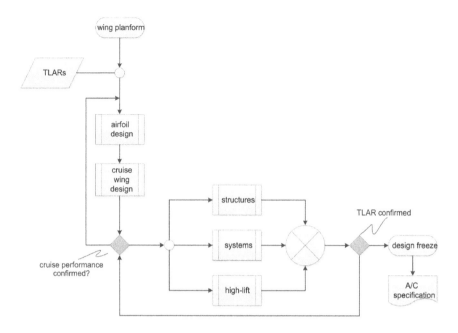

**FIGURE 8.2**   Flow chart of specific design phase of aircraft wing design.

the designs are merged and conflicts between the solutions of different disciplines are analyzed remembering that the final goal is compliance with the TLARs. The major conflicts arise from space allocations at the interfaces. For example, the wing spar positions defined by the structural design are crucial for the available space of systems and high-lift devices and the available volume for fuel storage. These conflicts are solved by running through several design loops and placing the requirements of a certain discipline as a side constraint into the design of the others. It is not uncommon that detected conflicts cannot be fully resolved within the initially given wing shape. In such cases, adoptions to the wing shape may be required to allow for more space to resolve the issues sufficiently for all disciplines for achieving the TLARs.

Looking into the design of the high-lift devices themselves, shown in Figure 8.3, the design process starts with the selection of the high-lift system layout. This

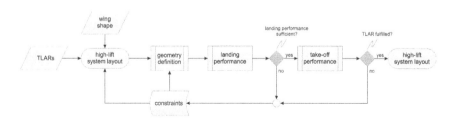

**FIGURE 8.3**   Flow chart of high-lift design process.

includes the pre-selection of the types of high-lift devices and the coverage of the wing in the limits given the wing shape and the wing planform, where also the primary flight controls are already defined and limit the available spanwise extent, especially at the trailing edge.

The next step to be taken carefully is the geometry definition of the high-lift devices. This includes the shapes and designated positions in a deflected position. The first design loop is targeted to achieve the required landing performance. The rationale for this is simply that the landing position is the one with the largest deflection. Take-off and approach settings will be defined later on in between the landing setting and the retracted position.

At the end, the target is a well-balanced high-lift system that approximates with the intermediate setting a continuous envelope with some flexibility in overlapping design ranges. The aircraft can then be adapted to the specific flight condition by adopting the high-lift system deflection. Figure 8.4 shows the polars of a designed transport aircraft for multiple settings for landing, approach, and take-off compared to the clean configuration. In the graph included are the values for the minimum selectable speed for the different settings, given by the minimum safety factors discussed in Chapter 3. A well-balanced high-lift system achieves a smooth distribution of the maximum lift coefficients for good speed flexibility. The shown lift vs. drag curves should also show a good approximation of the envelope that would be obtained by continuous movement. It should be remembered that the drag coefficient is only of importance at lift coefficients lower than the indicated values relating to the minimum speed as the upper part of the polar is not used in flight and therefore not relevant for fuel burn.

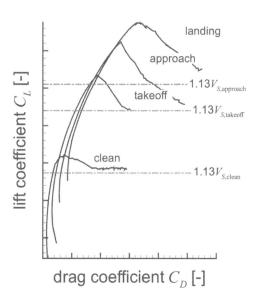

**FIGURE 8.4** Balancing of aerodynamic behavior for multiple settings of high-lift devices at the DLR-F11 high-lift configuration. [3]

## 8.2 GEOMETRIC DESCRIPTION OF HIGH-LIFT SYSTEMS

It is evident that, prior to any design, the degrees of freedom need to be specified. Since the high-lift devices are retracted for most of the flight time, any disturbance of the cruise wing shape may imply detrimental effects on cruise performance and must be avoided. The proper selection of the degrees of freedom should, therefore, consider as most as possible the restrictions imposed by other disciplines in the design of the aircraft.

### 8.2.1 HIGH-LIFT SYSTEM LAYOUTS

The first step in high-lift design is the specification of a wing layout and the type of devices. The proper pre-selection can be crucial for a "good" high-lift system. The layout has to conform with the definition of primary control surfaces at the wing, mainly the ailerons for roll control. Figure 8.5 shows three different layouts of high-lift systems of current aircraft widely in use. Although only three, they show the whole range of classical layout decisions.

First, it can be seen that the leading-edge system, either Krueger flap or slat, covers the full span of the wing. In case of a mixture of systems, it is obviously advantageous to place the less effective inboards as the lift coefficient requirement is relaxed close to the fuselage. For very large aircraft, the most inboard part of the wing can even be left out. Further, leading-edge devices are commonly interrupted at the engine position to allow a closer integration of the engine[1].

At the trailing edge, the roll control surfaces dictate the spanwise extension of the flap system. Depending on the torsional stiffness of the wing and the desired cruising speed, an inversion of the roll control function has to be avoided. For the shown example of the Boeing 747, the rolling functionality at high speed is achieved by a high-speed aileron that is placed very inboard just in the position of the inboard engine. The position is carefully selected to obtain a so-called thrust-gate, as if the "engine jet would impinge on the flap, causing high flap loads and undesirable powered lift effects" – (Rudolph [1] p. 139). At low-speed condition, the roll

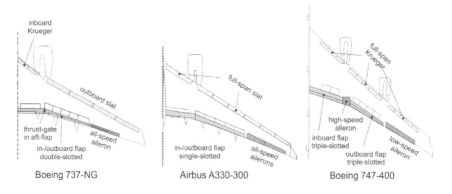

**FIGURE 8.5** Different layouts of high-lift devices of current civil passenger aircraft. (Wing Top Views Extracted from [4, 5].)

control is achieved by the aileron in the classical outboard position. The wings of the Airbus A330 and the Boeing 737-NG do not show the high-speed aileron as the roll function is supported by spoiler deflections, making the inboard aileron unnecessary. Nevertheless, the Boeing 737-NG shows a thrust-gate as the interruption of spanwise continuity of the aft-flap of the double-slotted Fowler flap system. These thrust-gates may be avoided for single-slotted flaps, as common on all Airbus aircraft past the A320 and seen for the A330 in the figure. Rudolph [1] states that "the biggest detriment to trailing-edge-flap performance seems to be spanwise discontinuities created by thrust gates and/or inboard, high-speed ailerons. These discontinuities are, of course, more severe for flaps with increased numbers of elements" – (Rudolph [1] p. 98).

### 8.2.2  PLANFORM PARAMETERS

The planform definition of high lift devices includes the specification of the spanwise and lateral extent of the devices seen in a top view. Although the overall maximum extent is mostly given by the wing layout, these limits are to be seen as upper limits as a reduction of the size of high-lift systems – if possible, in terms of aerodynamic performance – offers a potential of weight reduction. Further on, only the maximum spanwise extent is defined by the wing layout but not the subdivision into the several elements. Since the later shape and setting design orientates at the edges of the devices, the proper placement is an aid to achieve uniform settings along the span, especially for bent and twisted wings. Placing the design sections at the edges of devices hereby allows ensuring a continuous high-lift system.

In lateral direction at each design section, two degrees of freedom are identified, although the leading and trailing edges of the cruise wing are fixed. For a slat, there is the slat trailing edge itself, but also the forward position of the leading edge of the remaining main wing is free to design. For a flap, the flap size is similarly a degree of freedom as the trailing edge of the main wing forms the shroud cover and is usually used to place the spoilers. For Krueger flaps, it is the front and rear position of the wing cuts specifying the panel size. In any case, the planform includes two parameters per design section.

On tapered wings, the propagation of the planform parameters along the span can follow two different philosophies. In a constant chord design, the high-lift devices get the same chordwise extent, thus being not tapered. This increases the relative high-lift device size towards the wing tip, where, for the loading distribution, a higher maximum lift coefficient is needed than, e.g., at the wing root. Alternatively, in a constant relative chord design, the width of the high-lift devices scales with the local wing chord of the tapered wing. In such designs, an isobaric concept can be followed to obtain self-similar flow at the different wing stations. Such a design can have benefits especially in drag coefficient at high angles of attack but rejects last percentages in lift generation. Therefore, there is no clear preference for the one or other strategy, as different successful aircraft demonstrate and the best design is largely interdependent with the architecture of the whole wing.

At slats, there is a distinct trade-off of aerodynamic performance and manufacturing effort. "The more-or-less constant-chord slats on most Boeing airplanes are probably optimum for cost reduction but not for high-lift performance. [...] The tapered slats found on the Douglas airplanes probably are a good match for aerodynamic performance, but the slat mechanization has its drawbacks: First, the multitude of different radius tracks add to manufacturing cost" – (Rudolph [1] p. 85). This trade-off is identified to be related to the different types of motion needed for straight or tapered devices. "Cylindrical slat motion of nearly constant-chord slats allows the use of identical slat tracks and a simple actuation system for an overall savings in complexity and cost. Conical motion of tapered slats by today's standards requires that all slat tracks have a different radius and a complex actuation system with high manufacturing cost" – (Rudolph [1] p. 91).

For the inboard wing of the common double-trapezoidal wing planform of transonically operating transport aircraft, the chord strategy is more homogeneous throughout aircraft manufacturers. The spanwise wing loading does not require very high lift coefficients in this area and the chord increase provides by itself a large contribution to the wing area. For these reasons, the inboard wing most often incorporates constant (absolute) chord devices.

### 8.2.3   SHAPE PARAMETERS

The major requirement to not impact cruise performance implies that the shape forming the cruise wing has to be retained. This implies that only those parts of the shape of a high-lift wing can be altered that are hidden or retracted in cruise flight. For a state-of-the-art high-lift system with slats and flaps, this still gives a number of degrees of freedom. For a flap, it is the leading edge and upper side shape that is covered by the shroud in retracted position. For the slat devices, it is, at the end, not the slat shape that is designed, but the fixed leading edge of the main wing that is covered by the retracted slat.

For the shape definition, any kind of geometric function can be used for single airfoils. Nevertheless, in the past, a certain preference can be found on geometric representations based on splines. Of additional interest are conical curves, as they have a linear variation of curvature and, therefore, obtain smoothest pressure distributions. In any case, a strong effort should be taken to obtain curvature-steady shapes as any discontinuity in curvature results in distortions of the pressure distribution. At the very strong pressure gradients along the high-lift airfoil, such imperfections may generate artificial separations or separation bubbles.

### 8.2.4   POSITION PARAMETERS

The last set of degrees of freedom for design contains the positions of the high-lift systems relative to the wing. In case of classical slats and flaps, or similar, these are the degrees of freedom of the rigid body motion of deployment. For the two-dimensional wing section, these are the horizontal and vertical displacements together with the deployment angle.

The definition of the deployment angle is relatively unique and describes the rotation angle from the retracted to the deflected position. The position displacements may be described in different ways, outlined in Figure 8.6. Mathematically, easiest

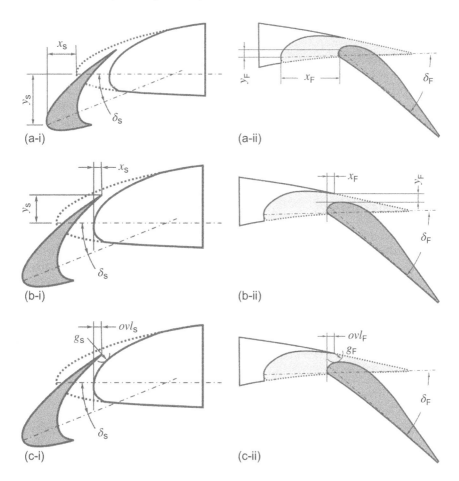

**FIGURE 8.6** Definitions of position parameters for placements of slat and flap devices: (a) mathematical; (b) Woodward & Lean [7]; (c) aerodynamic.

to implement is a definition based on a fixed coordinate of the device, e.g., the slat or flap leading edge. The drawback of such a definition is that it does not account for the sensitivity of the aerodynamics on the variation of the position. In Section 5.2, the sensitivities of slotted systems have been discussed in detail. The major observation has been the strong sensitivity to the gap width of the slot. Woodward and Lean [7] proposed a coordinate system using the trailing edge of the slat as origin. For the flap, they used the intersection of the horizontal and vertical tangents to the rotated flap as reference point. For the mathematical description, this is still relatively simple as the reference point only depends on the deployment angle and the calculation is limited to the identification of the minimum x-coordinate and maximum y-coordinate of the flap.

From an aerodynamic point of view, it is best to directly use the characteristic of the slot, namely the gap and overlap definition. The gap is defined as the minimum distance of the trailing edge of the preceding airfoil element to the surface of the downstream element. The overlap is the vertical projection of the trailing edge of the upstream element to the leading edge of the following element, which is for the flap at the vertical tangent to the deflected flap surface. The overlap is defined positive if there is a real overlap and negative if, when looked at from above, also a gap in horizontal direction is visible. The overlap definition is equivalent to the x-position by Woodward and Lean.

Regarding the three-dimensional wing, five parameters remain for a single element, since the deflection angle is the same, but the position of the wing can be adjusted at both ends. Nevertheless, it is, in most cases, favorable to build up a continuous device to avoid gaps in between the high-lift device elements. Therefore, if no interruption is foreseen, the deflection angle is only one parameter, and the number of stations for the displacement is the number of elements plus one.

## 8.3   CONSTRAINTS

In the context of design, a constraint is a side condition to be respected for the designs to be feasible. The simplest type of constraint is a range boundary on a design parameter. But constraints in aircraft design most often are defined by the interface to other disciplines within the design context. Such constraints are much harder to implement and to respect during design due to their non-linear nature. For high-lift design, major constraints arise from the cruise wing design and from structures and systems.

### 8.3.1   STRUCTURAL CONSTRAINTS

Structural constraints limit the degrees of freedom in design to guarantee structural integrity, stiffness, and strength of the aircraft. This does not only apply to the wing structure but also to the strength of the devices themselves.

The positions of the wing spars are a hard constraint for any high-lift design. They are the major load-carrying structural elements and their structural design is driving wing weight to a large extent. Similar to the cruise wing shape, any impact on the spar should be avoided. Even though it seems feasible to overlap the high-lift device extent with the spar, e.g., at the slat trailing edge, such overlapping would reduce the spar height and is not desired. But also, the structural integrity of the high-lift devices themselves must be considered. While slats are short and relatively thick, the flaps can get quite flat and a minimum thickness or spar height within the flap must be respected. Otherwise, the flap weight will get unacceptably high.

A specific area of interest is the trailing edge of an upstream element of a multi-element wing. For the downstream element to be smooth at the upper side, this upstream trailing edge ideally would be tangential at no thickness – and, therefore, without any structural stiffness. On the other hand, the upstream trailing edge should be stiff avoiding a bending under aerodynamic loads away from the downstream element in order to guarantee the desired gap values. This requires, first, a certain thickness of the trailing edge and a required opening angle between upper and lower sides to achieve the required bending stiffness. Figure 8.7 illustrates the corresponding constraints at the slat and the wing trailing edge. Typical values are – depending on the aircraft size – trailing edge thicknesses in the range of $d_{TE} = 2$ mm ÷ 5 mm and opening angles in between $\gamma_{TE} = 3° ÷ 5°$.

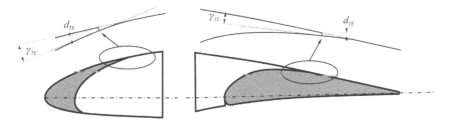

**FIGURE 8.7**   Structural constraints regarding the trailing edge of the upstream element of multi-element airfoils.

The consequence of the finite thickness is a local disturbance to the smoothness of the upper side of the downstream element as well as the clean wing shape. If a smooth clean wing shape should be retained, this would call for a forward-facing step in the shape of the deployed high-lift configuration. Avoiding this leads to a backward-facing step in the clean wing shape. This is the special reason why slats are not useful as a high-lift device for a wing that is incorporating laminar flow technology. At the high-lift shape, still a discontinuity remains. As a sharp edge should be avoided to mitigate any risk of flow separation, this continuity is usually smoothed out. Figure 8.8 shows such a blending at the trailing edge of a slat device. The smooth transition from the high-lift surface to the clean wing shape results in a local curvature increase leading to a local suction peak even visible in the pressure distribution. The curvature increase is thereby increasing with reduced trailing edge thickness

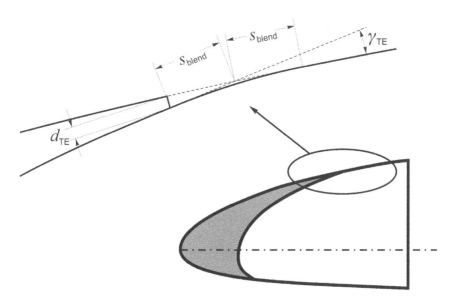

**FIGURE 8.8**   Blending of inner high-lift contour into clean wing shape at the slat trailing edge.

and increased opening angle. The minimum distance needed for a smooth transition is estimated by

$$s_{blend} = \frac{d_{TE}}{\tan \gamma_{TE}}.$$    (8.1)

Shorter blending distances will likely lead to inflections in curvature.

## 8.3.2  KINEMATICS CONSTRAINTS

In principle, the deflected positions could be defined on a pure aerodynamic perspective and the mechanisms for the positioning were attempted to be designed afterward. Although different mechanisms may be suitable for achieving one defined landing position, finding a mechanism that moves the device into three to five additionally defined positions along the path to the landing position gets difficult. The right choice of the mechanism, or kinematics, is crucial for a proper deflection of the high-lift systems. It is, therefore, worth to account for the capabilities of a kinematics type already from the beginning.

At the trailing edge, the flap is exposed to substantial aerodynamic loads, which the kinematics has to transfer into the wing. The common risk is therefore that an ideal kinematics that enables the best positions from an aerodynamic point of view may get too heavy and too complicated to be applicable for an aircraft. The latter additionally implies an increased effort for maintenance and overhaul. There is a large number of patents on kinematics around. Nevertheless, the most common ones are, according to Rudolph [1]: (i) fixed hinge; (ii) four-bar linkages; (iii) hooked track support; (iv) link/track kinematics.

The fixed hinge kinematics, shown in Figure 8.9, is the easiest kinematics with the least complexity. Only one bearing is needed per flap track station. This kinematics is feasible both for slotted flaps (see Section 5.2.4) and Fowler type 2 flaps (see Section 5.2.5). The flap performs a rotational or a helical motion in three dimensions. The hinge position is defined by the retracted and the most deflected position. Take-off settings must be placed on the motion path. A drawback of fixed hinge kinematics for Fowler flaps is the considerable length of the support struts. The length increases with increasing Fowler motion and shallower flap angles and can be reduced by an increased flap gap. Fixed hinge kinematics have mostly been used in conjunction with double slotted vane-flaps, either fixed vane (Douglas DC-9, Airbus A400M) or articulated vanes (Douglas DC-10, Fokker 100), where the vane gap is opened during deflection. The latter allows for a reduction on an undesirable large flap gap. A second method to reduce the gap again is by deflection of the spoiler forming the wing trailing edge downwards. If "all spoilers could be drooped for the landing flap positions [..] the flaps could be deflected to a higher angle. ... On a modern fly-by-wire airplane, the drooping of the spoilers should not cause a problem in terms of increased weight or systems complexity" – (Rudolph [1] p. 139). The method to droop the spoiler to control the flap gap has been first introduced by Douglas in the development of the C-17 Globemaster military transport aircraft [8]. On civil jet airliners, both Airbus and Boeing use fixed hinge flap supports in

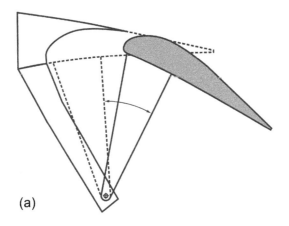

(a)

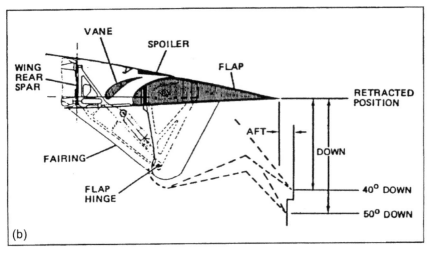

(b)

**FIGURE 8.9**  Fixed (dropped) hinge kinematics: (a) principle; (b) inboard flap kinematics of the Douglas DC-9. (From Rudolph [1], figure 1.12.)

conjunction with spoiler droop on the A350XWB [9] and the B787 Dreamliner [10], respectively.

A four-bar linkage is a system that can be reduced to a set of four bars linked together at the ends. One bar is the fixed connection to the wing. Pairwise equal lengths of such a kinematics would build a parallelogram with the ability to move the flap without deflection angle. The proper tuning of the lengths allows building up a movement where the flap is first at low deflection angles displaced to the rear for wing area increase and only rotated at the last stage of the deflection path. Such systems are common on the Boeing B757, B767, and B777. Figure 8.10 shows the kinematics of the B767.

Hooked track supports, as shown in Figure 8.11, were used on the earlier Boeing airplanes B707, B727, B737, and B747 and earlier Airbus airplanes A300 and A310.

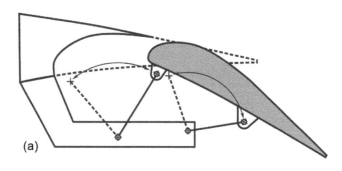

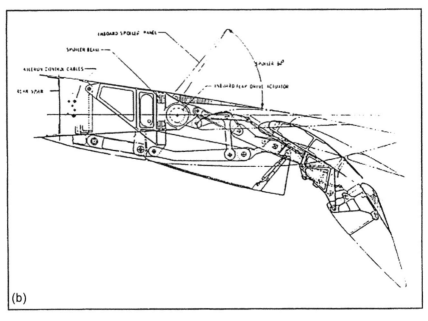

**FIGURE 8.10**  Four-bar linkage: (a) principle; (b) inboard flap kinematics of the Boeing 767. (From Rudolph [1], figure 1.29.)

The flap is mounted on a carriage that runs on a track that is bent at the end. The big advantage is the very high degree of freedom to design the path of the flap deflection. The big disadvantages are the high weight and manufacturing cost of the tracks. The manufacturing tolerances have to be very tight at a very high accuracy. On the other hand, the guided rollers on both sides are impacted by a high wear and tear.

Airbus introduced link/track mechanisms, illustrated in Figure 8.12, on the A320 and kept them up to the A380. In contrast to the hooked track support, the carriage runs on a straight track that is much easier to manufacture. The rotation of the flap is achieved similarly to the four-bar linkage by introducing a front or rear link.

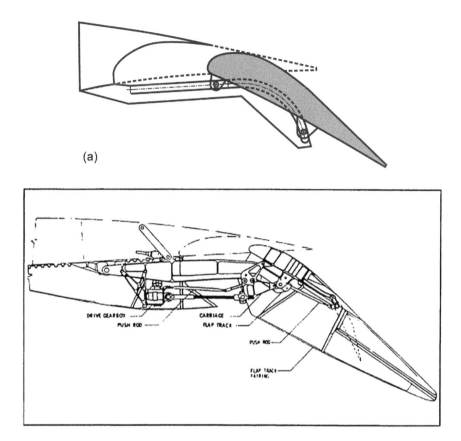

(a)

**FIGURE 8.11**    Hooked track support: (a) principle; (b) flap support of the B757 inboard flap. (From Rudolph [1], figure 1.30.)

Rudolph [1] summarized the deflection paths of the different types of kinematics in Figure 8.13. The flap aft motion is shown in relation to the deflection angle. The intention is a large movement and a late rotation. "It appears that all three link/track mechanisms are better than either the 757 hooked track or the 777 four-bar linkage in producing high-takeoff Fowler motion at small flap angles. The 767 complex, four-bar linkage also develops high Fowler motion at low flap angles" – (Rudolph [7], p. 113). One specific characteristic can be seen at low deflection angles. At the beginning of the deployment, the flap can have a negative deflection angle. This is intended as it helps the flap to slide below the spoiler. Within the first half of the path, the flap can do not more than a 5° deflection. This reduces the drag in the take-off setting, as the main lift increase is governed by the wing area increase.

The typical mechanism to deploy a slat is a movement on a circular arc track. This kind of mechanism has been introduced by the Douglas Aircraft Company on the

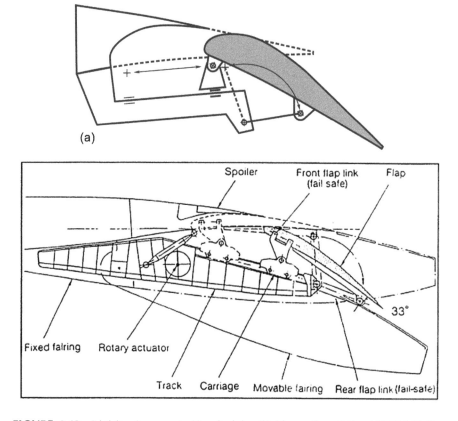

**FIGURE 8.12** Link/track support: (a) principle; (b) kinematics of the A330/A340 flap. (From Rudolph [1], figure 1.34.)

DC-8 aircraft. Today, most large aircraft – Airbus, Boeing, Bombardier, Embraer, etc. – are equipped with slats on such kind of a kinematics. The slat is mounted on circular arc tracks that are guided by rollers, as shown in Figure 8.14 for the Boeing B757. A pinion drive steers the deployment, positioning the slat in a forward and downward inclined position. An important aspect of the mechanical design is the length and the radius of the circular arc track. It usually penetrated the front spar, and the fuel tank behind must be sealed against it. Further on, the slat track may not be as long that it additionally penetrates the lower wing surface in retracted position.

The provision of the gap "is well suited to achieve $C_{L,\max}$ targets for landing configurations, but is contradictive to keep the drag of a takeoff configuration low. Also the noise emission of a vented leading edge device is significantly higher than for an un-slotted solution" – (Strüber [11], p. 2). The evolution of the slat gap with the deflection angle is defined by the shape of the fixed leading edge of the main wing. "Although with careful optimization a low drag level can be obtained with a conventional vented slat, a sealed slat […] [can] be even more drag efficient" –

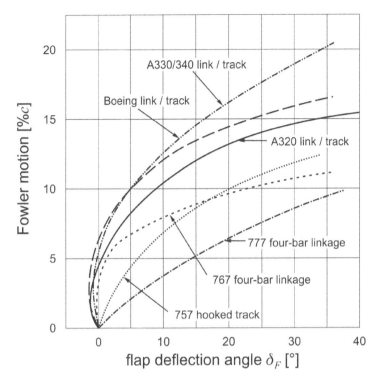

**FIGURE 8.13** Deflection paths of different types of flap support kinematics. (According to Rudolph [1], figure 3.21.)

(Strüber [11], p. 4). This sealing at low deflection angles can be achieved by a leading-edge contour following the deployment path of the slat trailing edge, as, e.g., implemented at Boeing 777 [12] or the Airbus A350-900 XWB [11]. An alternative is the provision of a secondary set of "slave links that run in programming tracks and rotate the slat panel counter to the rotation provided by the circular arc track, which means that the slat is attached to the main tracks with only one pin in each location to allow for this rotation" – (Rudolph [1] p. 10). Such programming tracks have been implemented, e.g., at the Douglas DC-9 and DC-10 or the Boeing 757 and 767 [1].

## 8.4   RELATION OF AIRFOIL AND WING DESIGN

Usually, it is advantageous to first design a proper high-lift wing airfoil section before caring about the three-dimensional effects of the full wing. The simple reason for this is the required design time. Using numerical simulation, the effort by neglecting the third dimension not only reduces the number of computational points but also reduces the number of equations to be solved. Nevertheless, in order to properly design for the conditions at the real wing, it is necessary to carefully select

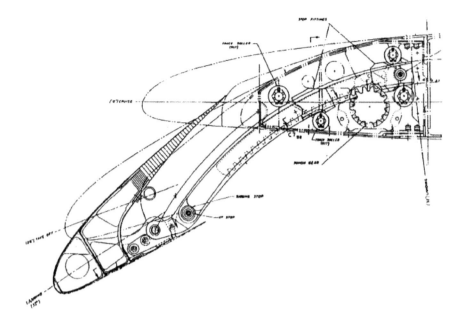

**FIGURE 8.14**   Rack & pinion mechanism to deploy a slat, here from the Boeing 757 aircraft. (From Rudolph [1], figure 1.21.)

an appropriate wing section for design and to apply a correct forward transformation of the geometry and flow conditions as well as the appropriate backward transformation of the aerodynamic properties.

### 8.4.1   WING DESIGN SECTION

The selection of the appropriate wing section for design is driven by the location of the most highly loaded position along the wing span, as this section will most likely trigger stall onset. At this point, it is necessary to understand the difference between the lift distribution and the lift coefficient distribution. Figure 8.15 shows a comparison of both for a clean wing without high-lift devices for different angles

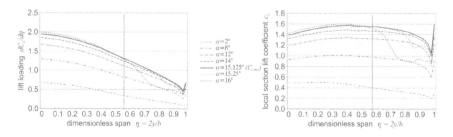

**FIGURE 8.15**   Comparison of (left) lift distribution and (right) lift coefficient distribution for different angles indicating stall onset.

of attack. The lift distribution represents the spanwise derivative of the overall lift coefficient and directly scales with the local circulation

$$\frac{\partial C_L}{\partial \eta} = \frac{2}{U_\infty c_{ref}} \frac{\partial \Gamma}{\partial \eta}.$$ (8.2)

It is, therefore, equivalent to the circulation distribution, which is optimal for induced drag if its shape is elliptical. The lift coefficient distribution gives the indication of the loading of the local wing section in relation to its lift potential by showing the local lift in airfoil dimensions that relate to the local wing chord. Both distributions can be converted by the simple geometric relation

$$c_L(\eta) = \frac{\partial C_L}{\partial \eta} \frac{c_{ref}}{c(\eta)}.$$ (8.3)

Especially for the maximum lift design, the wing section for design shall be taken from the most loaded portion of the wing. This area does not only depend on the planform of the wing but also on the wing twist distribution. Usually, the design wing section of a typical transport aircraft wing is located at a dimensional span of $\eta = 55 \div 65\%$.

## 8.4.2 SWEPT WING TRANSFORMATION

The wings of civil transport aircraft are usually swept to achieve high cruising speeds. According to Busemann [13], only the flow perpendicular to the surface curvature impacts the pressure distribution and thereby the aerodynamic forces. Although the theory has been developed for the transonic cruise regime, it applies to the low-speed flight in the same way. It is, therefore, necessary to transform the design wing section using the wing sweep and not to analyze the geometrical wing cut in the line of flight. The flow around the wing section is thereby approximated by the flow around an infinite swept wing with the same sectional airfoil. Since the usual aircraft wing is additionally tapered, the wing sweep is not constant along the airfoil chord. This makes the selection of the wing sweep to apply a bit crucial and the choice to use the leading-edge sweep for high-lift design as shown in Figure 8.16 will be motivated later in this section.

In Busemann's swept wing theory, the flow is divided into the two-dimensional flow perpendicular to the wing sweep with the magnitude of the onflow velocity

$$U_{\infty,2D} = U_{\infty,3D} \cos \varphi$$ (8.4)

and the one-dimensional flow along the swept wing span, the latter being constant

$$w_\infty = U_{\infty,3D} \sin \varphi.$$ (8.5)

Geometrically, the swept wing transformation refers to a compression of the wing section in chord direction. As the airfoil chord reduces while the thickness stays

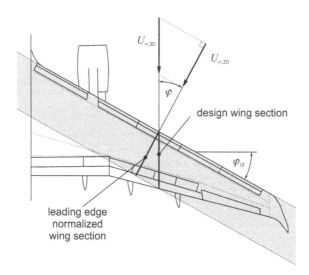

**FIGURE 8.16**   Swept wing transformation of wing design section airfoil.

constant, the airfoil becomes relatively thicker. Since airfoil analysis usually uses the chord length for non-dimensionalization

$$
\begin{pmatrix} \tilde{x} \\ \tilde{y} \end{pmatrix} = \frac{1}{c} \begin{pmatrix} x \\ y \end{pmatrix},
$$

(8.6)

the constant sweep transformation results in a thickening of the airfoil scaled to unity chord

$$
\begin{pmatrix} \tilde{x}_{2D} \\ \tilde{y}_{2D} \end{pmatrix} = \frac{1}{c} \begin{pmatrix} x_{3D} \\ \dfrac{1}{\cos\varphi} z_{3D} \end{pmatrix}.
$$

(8.7)

It is necessary to respect that the axis of rotation is not perpendicular to the two-dimensional plane. Therefore, the effective sweep angle is equal to the geometric sweep angle and maximum only at $\alpha = 0°$. Imagine a 90° incidence, the effective sweep angle reduces to zero

$$
\cos\varphi_{eff} = \cos\varphi \cdot \sqrt{1 + \sin^2\alpha_{3D}\tan^2\varphi}.
$$

(8.8)

The transformation of the flow speed and the geometry leads to the adaptation of the characteristic flow parameters

$$
M_{\infty,2D} = \cos\varphi_{eff} \cdot M_{\infty,3D}
$$
$$
Re_{\infty,2D} = \cos^2\varphi_{eff} \cdot Re_{\infty,3D}.
$$

(8.9)

The projected angle of attack results from the geometric projection of the onflow direction onto the two-dimensional flow plane.

$$\tan \alpha_{2D} = \frac{\tan \alpha_{3D}}{\cos \varphi_{eff}}. \tag{8.10}$$

It is obvious that the scaling of the Mach number mainly affects the corresponding consideration of compressibility effects. For tapered wings in cruise conditions, the reference sweep angle is, therefore, often taken at the mid chord to properly capture the effect of the transonic shock. For high-lift systems, the effect of compressibility mainly takes place close to the leading edge. Therefore, for high-lift system aerodynamics, the leading-edge sweep is more representative. Thus, the scaling is the so-called leading-edge normalization.

The backward transformation scales the fluid state back to the three-dimensional conditions. Since this scaling only transforms back to the infinite swept wing, this is often denoted as the 2.5D state. Since the definition of the pressure coefficient scales with the square of the onflow Mach number or velocity, the pressure coefficient for the 2D airfoil section relates to the pressure coefficient at the infinite swept wing as

$$c_{p,2.5D} = \cos^2 \varphi_{eff} \cdot c_{p,2D}. \tag{8.11}$$

Neglecting any contribution by friction to the lift coefficient, its relation is analog

$$c_{L,2.5D} = \cos^2 \varphi_{eff} \cdot c_{L,2D}. \tag{8.12}$$

The pressure force of the airfoil section contributes to a spanwise force component, leaving only the projected part in the line of flight. The pressure part of the drag, therefore, scales by

$$c_{Dp,2.5D} = \cos^3 \varphi_{eff} \cdot c_{Dp,2D}. \tag{8.13}$$

On the other hand, the boundary edge velocity at the infinite swept wing is higher than for the airfoil section by the component along the wing span. An estimate to scale the friction part of the drag is thereby the scaling of the wall shear stress leading to

$$c_{Df,2.5D} = \cos \varphi_{eff} \cdot c_{Df,2D}. \tag{8.14}$$

The aerodynamic characteristics of the infinite swept wing can be further used in lifting-line methods (see Section 7.2.3) for design. The last step of the transformation back to the full wing needs some pre-knowledge about the three-dimensional flow and is mainly suited for incremental design, namely the improvement of an already existing design. Given an initial configuration and information about the flow, both on the full 3D wing and the corresponding design section airfoil allows establishing a relation to predict the full wing behavior based on two-dimensional airfoil data. In a first step, it is necessary to correlate corresponding flow conditions, as the angle of

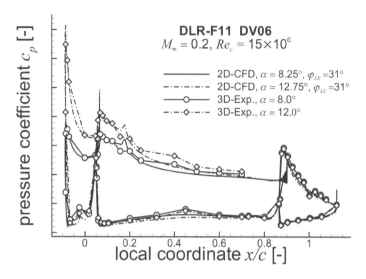

**FIGURE 8.17** Comparison of pressure distributions of the full three-dimensional wing and the corresponding leading-edge normalized two-dimensional infinite swept wing flow. (From Wild et al. [3], figure 7.)

attack of the wing section of a full three-dimensional wing is further influenced by the local wing twist and the induced angle of attack due to spanwise variations of the circulation distribution. The proper (and only) way to achieve this correlation is by matching the pressure distributions for both cases. Figure 8.17 shows such a pressure distribution matching between the infinite swept wing obtained by the leading-edge normalized wing section achieved by CFD with the pressure distribution of the full three-dimensional wing obtained from wind-tunnel measurements. The deviation in the angle of attack represents the mentioned offset induced by the wing twist and induced incidence.

Having obtained the matching pressure distributions, both represent the same flow condition. From the two- and three-dimensional data, the correlation

$$C_{L,3D} = \gamma_{C_L} C_{L,2.5D} \tag{8.15}$$

can be derived assuming that changes in the sectional lift coefficient only affect the level of the lift-loading and not the shape of the spanwise lift distribution. For the drag coefficient, a similar relation applies with the extension to account for the change of the induced drag due to changes in the lift coefficient

$$C_{D,3D} = \gamma_{C_D} \left( c_{Dp,2.5D} + c_{Df,2.5D} \right) + \frac{\left( \gamma_{C_L} C_{L,2.5D} \right)^2}{\pi \Lambda} \tag{8.16}$$

The correlation factors $\gamma_{C_L}$ and $\gamma_{C_D}$ are obtained from the values at matching pressure distributions. The validity of the back-scaling to the three-dimensional wing is

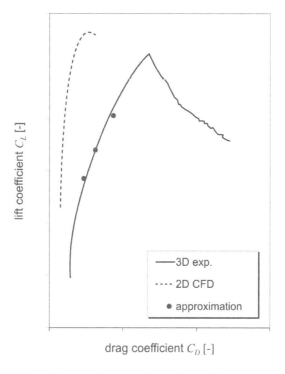

y-axis: lift coefficient $C_L$ [-]

x-axis: drag coefficient $C_D$ [-]

Legend:
——— 3D exp.
- - - - 2D CFD
● approximation

**FIGURE 8.18** Validity of swept-wing transformation scaling to represent three-dimensional wing aerodynamics by two-dimensional wing section analysis.

shown in Figure 8.18 where the polars are shown for the two-dimensional wing section from CFD, the three-dimensional wing data from experiments, and the results of the above approximation.

### 8.4.3 LOCAL SWEEP TRANSFORMATION

Wings of transport aircraft cruising at transonic speed incorporate wing sweep reducing the wave drag and an additional taper to reduce induced drag. Therefore, leading and trailing edge sweeps are not equal. The local sweep transformation attempts to reproduce the thickness and curvature distribution of the tapered and swept wing on a two-dimensional wing section. Analog to the swept wing theory, a design section from the wing is chosen. In the local sweep transformation, the wing is then represented by a conical surface, as illustrated in Figure 8.19, which is defined by the leading and trailing edge sweep of the wing [14].

The developed two-dimensional conic wing section is obtained as a cut through the conical wing at constant radius, which is defined by

$$r = \frac{c}{\sin \varphi_{LE} - \cos \varphi_{LE} \tan \varphi_{TE}}. \tag{8.17}$$

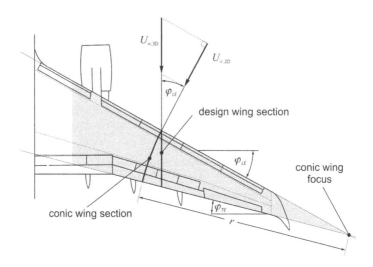

**FIGURE 8.19** Local sweep transformation of wing design section airfoil.

The coordinates of the center of the conic surface in the coordinate system of the wing design section are, therefore,

$$\begin{pmatrix} x_c \\ y_c \end{pmatrix} = r \cdot \begin{pmatrix} \sin \varphi_{LE} \\ \cos \varphi_{LE} \end{pmatrix} \qquad (8.18)$$

and the local sweep of a conical line passing through a point of the wing design section is

$$\tan \varphi(x) = \frac{x_c - x}{y_c}. \qquad (8.19)$$

Since the radius of the points along the generating wing section towards the conic surface center reduces, the conic wing section gets thicker as it approaches the trailing edge

$$z(x) = z \frac{\cos \varphi(x)}{\cos \varphi_{LE}}. \qquad (8.20)$$

As airfoil aerodynamics are related to the sectional chord length, which is for the conic wing section

$$c = r \cdot (\varphi_{LE} - \varphi_{TE}), \qquad (8.21)$$

the full transformation of the coordinates is given by

$$\begin{pmatrix} x \\ y \end{pmatrix}_{conic} = \frac{1}{r(\varphi_{LE} - \varphi_{TE})} \begin{pmatrix} r \cdot (\varphi_{LE} - \varphi(x)) \\ z \dfrac{\cos \varphi(x)}{\cos \varphi_{LE}} \end{pmatrix}. \qquad (8.22)$$

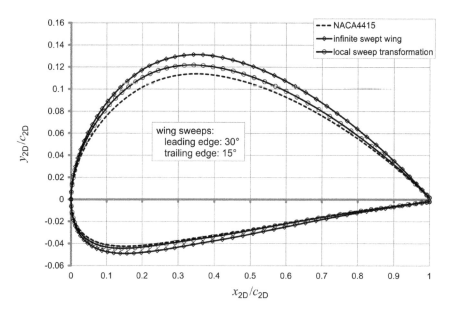

**FIGURE 8.20** Comparison of transformed airfoil shapes from a NACA4412 wing with 30° leading edge sweep and 15° trailing edge sweep.

The local sweep transformation is often misinterpreted as applying the thickness scaling similar to the swept wing but with the local sweep angle. But both the infinite swept wing and the local sweep transformations are mainly mappings of the airfoil length – the thickening of the airfoil results in the normalization to the changed wing chord. It shall be highlighted here that this leads to a different airfoil. Figure 8.20 shows a comparison of the different transformations of a NACA 4412 wing section for a wing tapered with 30° and 15° at the leading and trailing edge, respectively.

The transformation of the pressure coefficient has been derived by Lock [15] to

$$c_{p,3D} = \frac{f-1}{\frac{1}{2}\gamma M_{\infty,3D}^2} + f \cdot c_{p,2D}\cos^2\left(\varphi_{eff}\right)$$

$$f = \frac{1+\frac{1}{2}(\gamma-1)M_{\infty,3D}^2\cos^2\left(\varphi(x)\right)}{1+\frac{1}{2}(\gamma-1)M_{\infty,3D}^2\cos^2\left(\varphi_{eff}\right)},$$

(8.23)

where the effective sweep angle is given by eq. (8.8) and it is the one that is used to scale the flow conditions. For the lift and drag coefficients, no easy backward transformation is given, and it is necessary to evaluate the line integral along the curved conic wing section.

Figure 8.21 shows a pressure distribution of a multi-element high-lift airfoil of a swept and tapered wing obtained by the local sweep transformation in comparison

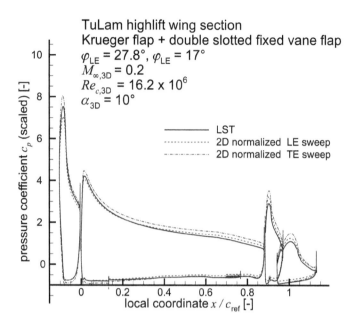

**FIGURE 8.21** Pressure distribution obtained by local sweep transformation in comparison to 2D simulations normalized with leading edge or trailing edge sweep.

to the ones obtained by 2D simulations normalized according to eq. (8.11) either with the leading edge or the trailing edge sweep. The comparison highlights the need for the LST for a more exact representation. This is seen either at the stagnation pressure levels, which can be reproduced by the LST for the varying sweep angle, or at the suction level. It is seen that in the LST the suction level on the leading-edge device is higher than predicted by the leading edge normalized swept wing analogy. This is due to the increased upstream effects. Additionally, the suction level on the rear part of the multi-element airfoil falls below even the level of both 2D-normalized simulations. So due to the downstream effects, the flap flow is even more relieved although the stagnation pressure is increased.

### 8.4.4 Transformations of High-Lift Device Deflections

High-lift system devices are orientated along the leading and trailing edges of the wing. A similar gap and overlap can be obtained by displacement either in line of flight or perpendicular to the edge. Though, the rotational degree of freedom is to be defined in an axis that is somehow parallel to the respective edge. The orientation follows hereby to some extent structural requirements.

Leading edge devices are usually deployed normally to the front spar. The wing ribs supporting the kinematics are more easily designed and manufactured being perpendicular to it. As those systems mostly perform a rotational movement, the displacement direction is orthogonal to the front spar and the rotation axis is almost parallel. For this reason, leading edge deflections usually are defined in a coordinate system normal to the front spar.

Trailing edge devices on the inboard wing of swept back wings need a displacement in line of flight, as the flap would otherwise interfere with the fuselage. This also requires a similar movement of the outboard flap if no thrust gate is implemented. It is, therefore, straightforward to define a flap setting in a coordinate system in line of flight. Anyhow, the flap's rotation needs to be performed along a swept axis to achieve a homogenous gap and overlap setting along the span. The corresponding reference is, therefore, the trailing edge of the remaining main wing, e.g., the spoiler trailing edge.

Both definitions imply that the wing sections of a deflected high-lift system in line of flight do not comply with the wing sections in retracted position. In general, the sectional cuts are stretched and, therefore, deflected at swallower angles. Similar to the swept wing theory above, the deflected wing section is obtained by geometric transformations.

A series of transformations are mathematically more easily defined by harmonic coordinates. Here, an additional component is added to the coordinate vectors. In a two-dimensional problem, the coordinate vector is then

$$\mathbf{x} = \begin{pmatrix} x \\ y \\ 1 \end{pmatrix}. \tag{8.24}$$

Harmonic coordinates allow a straightforward implementation of basic geometric transformations as matrices. To map the three-dimensional movement of a high-lift device into the corresponding wing section, the sequence of operations consists of the scaling of the geometry according to the sweep $\varphi$ of the respective reference axis

$$\mathbf{S} = \begin{bmatrix} \cos\varphi & 0 & 0 \\ 0 & 1 & 0 \\ 0 & 0 & 1 \end{bmatrix}. \tag{8.25}$$

The rotation by the deflection $\delta^2$ around the origin is defined by

$$\mathbf{R} = \begin{bmatrix} \cos\delta & -\sin\delta & 0 \\ \sin\delta & \cos\delta & 0 \\ 0 & 0 & 1 \end{bmatrix}. \tag{8.26}$$

Finally, the translation

$$\mathbf{T} = \begin{bmatrix} 1 & 0 & \Delta x \\ 0 & 1 & \Delta y \\ 0 & 0 & 1 \end{bmatrix} \tag{8.27}$$

is applied to place the device in the right position. The full transformation is then given as

$$
\begin{pmatrix} x \\ y \\ 1 \end{pmatrix}_{deflected} = \mathbf{S}^{-1}\mathbf{TRS} \begin{pmatrix} x \\ y \\ 1 \end{pmatrix}_{retracted}
\tag{8.28}
$$

with the inverse of the non-uniform scaling to achieve the wing section in the original coordinate system.

Some care has to be taken to the definition of gap and overlap as those values depend on the orientation of the respective coordinate system and probable relative referencing using the retracted chord length as reference, as usually done for wing section analysis. For example, for slats, gaps, and overlaps are usually given as percentages to the local wing chord but in the system normal to the front spar. This results in a gap in the line-of-flight system reduced by the factor $\cos\varphi_{FS}$. For flaps, gap and overlap are specified in the line of flight system. For the evaluation of the translation, the gap has to be increased by the factor $1/\cos\varphi_{TE}$. Additionally, the flap deflection must respect the effective higher deflection angle in the swept coordinate system

$$
\tan\delta_{F,swept} = \frac{\tan\delta_{F,\text{line-of-flight}}}{\cos\varphi_{TE}}.
\tag{8.29}
$$

## 8.5  HIGH-LIFT DESIGN OPTIMIZATION

Counting the numbers of degrees of freedom and analyzing the interdependency of constraints, it is clear that aerodynamic high-lift design is a highly multi-dimensional problem with multi-disciplinary constraints. Regarding the number of design targets, it is also multi-objective. It is impossible to fully solve the design problem in an iterative manner by manual sensitivity analysis, especially as the different parameters are so interrelated and interdependent. A mathematical method to solve such a problem is numerical optimization.

Numerical optimization iteratively solves a design problem by variation and analysis of distinct individuals of the design space. For aerodynamic design by numerical optimization, this implies that for each of these individual designs a complete set of aerodynamic data must be calculated for each targeted design condition.

Performing such a design on the complete aircraft in high-lift conditions for all settings and design targets would imply the calculation of some tens of full three-dimensional simulations per design individual. Even with today's computer power, such a design would take months, maybe years to complete.

Full three-dimensional high-lift system optimization is therefore still out of scope[3]. High-lift wing section airfoil design is not as computationally expensive and has been introduced around the millennium[4]. At the current state, it can deliver a fully optimized high-lift wing section within a two weeks' time frame.

## 8.5.1  Methodology of Numerical Optimization

Mathematically, numerical optimization is defined as the identification of the minimum of a multi-variate so-called objective function

$$F_{obj}(\mathbf{x}) \mapsto \min.; \mathbf{x} \in \mathfrak{R}^n. \tag{8.30}$$

Here, $\mathbf{x}$ is the n-dimensional vector of design variables, which contains the degrees of freedom or design parameters. The objective function is minimal if and only if the gradient of the function vanishes

$$\nabla F_{obj}(\mathbf{x}_{opt}) = \mathbf{0} \tag{8.31}$$

and the curvature of the function is that the objective function increases in any direction away from the optimum, seen in a positive definite Hessian

$$\nabla^2 F_{obj}(\mathbf{x}) = \begin{bmatrix} \dfrac{\partial^2 F_{obj}}{\partial x_1^2} & \dfrac{\partial^2 F_{obj}}{\partial x_1 \partial x_2} & \cdots & \dfrac{\partial^2 F_{obj}}{\partial x_1 \partial x_n} \\ \dfrac{\partial^2 F_{obj}}{\partial x_2 \partial x_1} & \dfrac{\partial^2 F_{obj}}{\partial x_2^2} & & \vdots \\ \vdots & & \ddots & \\ \dfrac{\partial^2 F_{obj}}{\partial x_n \partial x_1} & \cdots & & \dfrac{\partial^2 F_{obj}}{\partial x_n^2} \end{bmatrix}. \tag{8.32}$$

The design problem may be constrained by additional functionals

$$\begin{aligned} g_i(\mathbf{x}) &> 0; i = 1, k \\ h_j(\mathbf{x}) &= 0; j = 1, m \end{aligned} \tag{8.33}$$

that represent the inequality and equality constraints that limit the design space. While the equality constraints are always in place, inequality constraints are only active if they would be violated. Any range limitation on a degree of freedom is equivalent to inequality constraints. Now, any equality constraint and active inequality constraint limit the design space in some direction, thus reducing the dimension of the design space by one each. When setting up a design problem for numerical optimization, it is easily achieved to overwhelm with constraint formulation on even low-dimensional design spaces. But to be able to achieve an optimum, the number of design variables must be greater than the number of active constraints. Otherwise, such a problem is ill-posed and can only be solved if the active constraints are not independent. The latter implies that some constraints could be eliminated without changing the design problem. If the constraints are independently active, then only the increase of variability by increasing the number of degrees of freedom offers the ability to retain a well-posed design problem.

The methods and algorithms to solve a numerical optimization problem are well described in several textbooks, e.g., Gill, Murry, and Wright [21], and numerous articles. Even today, new or adapted optimization methods get developed. The following is a comprehensive summary prepared earlier by the author [22].

From the mathematical formulation, the method of solving an optimization problem is a straightforward loop.

**Algorithm 8.1:** *Numerical optimization procedure*

---

$\mathbf{x} := \mathbf{x}_{ini}$

evaluate $F_{obj}(\mathbf{x})$

**Repeat**

      change $\mathbf{x}$

      evaluate $F_{obj}(\mathbf{x})$

**until** converged

$\mathbf{x}_{opt} := \mathbf{x}$

---

There are some characteristics that help to distinguish between the different types of algorithms. One major criterion is if the algorithm selects new sets of design variables strictly based on the sets already evaluated, as done in the so-called deterministic methods, or if the algorithm disturbs the design variables randomly. For the deterministic methods, a further classification is possible based on the use of the derivatives of the objective function. In order to build up a hierarchy of optimization algorithms, the starting point is the approximation of the objective function by its second-order Taylor series expansion around a set of design variables $\mathbf{x}$

$$F_{obj}(\mathbf{x}+d\mathbf{x}) = F_{obj}(\mathbf{x}) + \nabla F_{obj}(\mathbf{x}) \cdot d\mathbf{x} + \frac{1}{2} d\mathbf{x}^T \cdot \nabla^2 F_{obj}(\mathbf{x}) \cdot d\mathbf{x} + \mathcal{O}(3). \quad (8.34)$$

It is easily recognized that this approximation includes the two characteristics used for the optimality criterions, namely the gradient and the Hessian of the objective function.

### 8.5.1.1 Second Order Derivative Deterministic Methods

These methods use directly the second-order Taylor series expansion eq. (8.34). The gradient at a point $\mathbf{x}+d\mathbf{x}$ can be approximated by

$$\nabla F_{obj}(\mathbf{x}+d\mathbf{x}) = \nabla^2 F_{obj}(\mathbf{x}) \cdot d\mathbf{x} + \nabla F_{obj}(\mathbf{x}). \quad (8.35)$$

From the optimality condition eq. (8.31), the next local extremum is where the gradient vanishes. So, if the Hessian is known, the following system of linear equations has to be solved.

$$\nabla^2 F_{obj}(\mathbf{x}) \cdot d\mathbf{x} + \nabla F_{obj}(\mathbf{x}) = 0. \quad (8.36)$$

Since the objective function is normally not of the quadratic form, this procedure has to be iterated in order to obtain the minimum, leading to Newton's method.

**Algorithm 8.2:** *Newton's method*

---

$\mathbf{x} := \mathbf{x}_{ini}$

evaluate $F_{obj}(\mathbf{x})$

**Repeat**

       evaluate $\nabla F_{obj}(\mathbf{x})$

       evaluate $\nabla^2 F_{obj}(\mathbf{x})$

       solve $\nabla^2 F_{obj}(\mathbf{x}) \cdot d\mathbf{x} + \nabla F_{obj}(\mathbf{x}) = 0$

       $\mathbf{x} := \mathbf{x} + d\mathbf{x}$

**until** $\nabla F_{obj}(\mathbf{x}) = 0$

$\mathbf{x}_{opt} := \mathbf{x}$

---

The only prerequisite for finding a minimum is that the Hessian has to be positive definite. Otherwise, the actual guess is not within what is called the convergence region. For a stable algorithm, it is necessary to include the test of the Hessian to be positive definite and otherwise to do a steepest descent step.

One major shortcoming of the Newton method is that it requires the knowledge of the Hessian matrix. In typical applications, this is not the case and is only computable with high effort. Another solution is the use of the variable metric method, also called quasi-Newton method. The main idea of the variable metric method is to iteratively build up an approximation to the inverse Hessian.

$$\lim_{i \to \infty}(\mathbf{H}_i) \mapsto \left(\nabla^2 F_{obj}(\mathbf{x})\right)^{-1}. \tag{8.37}$$

In this case, the solution of the system of linear eq. (8.36) is replaced by the straightforward matrix-vector multiplication using the best approximation of the inverse Hessian

$$\mathbf{x}_{i+1} = \mathbf{x}_i + \mathbf{H}_i \nabla F_{obj}(\mathbf{x}_i). \tag{8.38}$$

The most common way for the update of the inverse Hessien is the Broyden-Fletcher-Goldfarb-Shanno (BFGS) update [23]. There is one positive side effect the BFGS update guarantees. If the initial inverse Hessian $\mathbf{H}_0$ matrix is positive definite, all subsequent updates $\mathbf{H}_i$ will be positive definite, too. Using the identity matrix as starting point may not be the best approximation, but it is positive definite and the first step of the algorithm will simply be a steepest descent step.

### 8.5.1.2 First Order Derivative Deterministic Methods

First order derivative methods, or so-called gradient methods, are used if the Hessian is too costly to be computed. They are then based on the Taylor series expansion of the objective function truncated after the first derivative

$$F_{obj}(\mathbf{x} + d\mathbf{x}) = F_{obj}(\mathbf{x}) + \nabla F_{obj}(\mathbf{x}) \cdot d\mathbf{x} + \mathcal{O}(2). \tag{8.39}$$

In order to reduce the objective function value, the dot product of the gradient $\nabla F_{obj}(\mathbf{x})$ and the perturbation vector $d\mathbf{x}$ has to be negative. This leads to methods, which define a search direction that holds negativity and perform a one-dimensional search in this direction. This so-called line search can again be looked at as a one-dimensional optimization problem.

**Algorithm 8.3: *Gradient method***

---

$\mathbf{x} := \mathbf{x}_{ini}$

evaluate $F_{obj}(\mathbf{x})$

**Repeat**

      evaluate $\nabla F_{obj}(\mathbf{x})$

      $d\mathbf{x} = f\left(\nabla F_{obj}(\mathbf{x})\right)$ with $\nabla F_{obj}(\mathbf{x}) \cdot d\mathbf{x} < 0$

      minimize $F_{obj}(\mathbf{x} + \alpha d\mathbf{x})$ for variable $\alpha$

      $\mathbf{x} := \mathbf{x} + \alpha d\mathbf{x}$

**until** $\nabla F_{obj}(\mathbf{x}) = 0$

$\mathbf{x}_{opt} := \mathbf{x}$

---

The steepest descent method is the simplest gradient method since it directly uses the negative gradient for the direction of search

$$d\mathbf{x} = -\nabla F_{obj}(\mathbf{x}). \tag{8.40}$$

From eq. (8.39), it can be derived that this direction improves the objective function mostly in the near region of the starting point. Nevertheless, if the design space has a curved shape the search direction is only the best in the near region of the starting point, while further away a search direction other than the negative gradient will yield better results. It can be shown that if the line search along $d\mathbf{x}$ has led to the local minimum in this direction, the gradient at this point will be orthogonal to the search direction

$$\nabla F_{obj}(\mathbf{x}_{i+1}) \cdot d\mathbf{x}_i = 0. \tag{8.41}$$

For a two-dimensional design space, this implies that only two alternating search directions will be used. In the most common case, for an arbitrary function, where the first gradient is not necessarily well aligned with the design space, this indicates that there are better ways of computing new search directions. The appropriate method for this is the use of the so-called conjugate gradient method. To explain this method, it is necessary to have a look again at the second-order Taylor series expansion eq. (8.34). For the search along an arbitrary direction $d\mathbf{x}$, it holds that the gradient may be approximated by

$$\nabla F_{obj}(\mathbf{x} + \alpha d\mathbf{x}) = \nabla F_{obj}(\mathbf{x}) + \alpha\left(\nabla^2 F_{obj}(\mathbf{x}) \cdot d\mathbf{x}\right). \tag{8.42}$$

Supposing that a local minimum has been reached along the search direction $d\mathbf{x}_i$, the new search direction $d\mathbf{x}_{i+1}$ should be chosen so that a motion along this direction does not spoil the minimization. This is the case if the gradients stay perpendicular to $d\mathbf{x}_i$, i.e., when the change in the gradient is perpendicular to $d\mathbf{x}_i$, leading to the conjugacy condition

$$d\mathbf{x}_{i+1}^T \cdot \nabla^2 F_{obj}(\mathbf{x}) \cdot d\mathbf{x}_i. \tag{8.43}$$

It can be proven that for a quadratic function $n$ line minimizations along a set of mutually conjugate directions lead exactly to the minimum, where $n$ is the dimension of the problem. Also, for functions that are not of the quadratic form repeated cycles of $n$ line minimizations will give quadratic convergence. Combining eq. (8.42) and eq. (8.43) leads to

$$\left(\nabla F_{obj}(\mathbf{x}_{i+1}) - \nabla F_{obj}(\mathbf{x}_i)\right) \cdot d\mathbf{x}_{i+1} = 0, \tag{8.44}$$

saying that the new search direction should be orthogonal to the change of the gradient, which is the case for

$$d\mathbf{x}_{i+1} = -\nabla F_{obj}(\mathbf{x}_{i+1}) + \frac{\left|\nabla F_{obj}(\mathbf{x}_{i+1})\right|^2}{\left|\nabla F_{obj}(\mathbf{x}_i)\right|^2} d\mathbf{x}_i. \tag{8.45}$$

The big advantage of the conjugate gradient method is that it accounts for the quadratic approximation of the objective function without a need to compute and store the Hessian matrix. The form of eq. (8.45) is the original Fletcher-Reeves version [24] which can be substituted as per a proposal of Polak and Ribiere [25]

$$d\mathbf{x}_{i+1} = -\nabla F_{obj}(\mathbf{x}_{i+1}) + \frac{\left(\nabla F_{obj}(\mathbf{x}_{i+1}) - \nabla F_{obj}(\mathbf{x}_i)\right) \cdot \nabla F_{obj}(\mathbf{x}_{i+1})}{\left|\nabla F_{obj}(\mathbf{x}_i)\right|^2} d\mathbf{x}_i. \tag{8.46}$$

Both forms are exactly equal for exact quadratic functions, but for real-world problems, where the quadratic form is not exact, the latter formulation delivers some convergence benefits.

### 8.5.1.3 Non-Derivative Deterministic Methods

If the computation of the gradients of the objective function is not applicable, neither directly nor per finite differences e.g., due to computational costs, a method is needed that solely relies on the objective function values of previously evaluated sets. Most algorithms of this type belong to the class of the quasi one-dimensional coordinate methods [21]. The various design variables are optimized independently resulting in loops over $n$ one-dimensional optimizations.

**Algorithm 8.4: *Coordinate method***

---

$\mathbf{x} := \mathbf{x}_{ini}$

**Repeat**

        **loop** over $i \in [1, n]$

                minimize $F_{obj}(\mathbf{x})$ for variable $x_i$

**until** $dF_{obj}(\mathbf{x}) = 0$

$\mathbf{x}_{opt} := \mathbf{x}$

---

Well-known examples of this type of method are interval reduction methods like the Golden Section algorithm. The prerequisite for these methods to work is that for each design variable an interval $\left[x_{i_{\min}}, x_{i_{\max}}\right]$ is given so that the desired minimum is encapsulated

$$F_{obj}\left(\mathbf{x}_{\min}\right) > F_{obj}\left(\mathbf{x}\right) < F_{obj}\left(\mathbf{x}_{\max}\right)$$

$$\mathbf{x}_{\min} = \begin{pmatrix} x_{1\min} \\ x_{2\min} \\ \vdots \\ x_{n\min} \end{pmatrix}, \quad \mathbf{x}_{\max} = \begin{pmatrix} x_{1\max} \\ x_{2\max} \\ \vdots \\ x_{n\max} \end{pmatrix}. \tag{8.47}$$

In a loop through the coordinate directions, they reduce the interval encapsulating the optimum down to a specified accuracy.

**Algorithm 8.5:** *Golden Section method*

---

$c_{gold} = \dfrac{\sqrt{5}-1}{2}$

$\mathbf{x}_1 = \mathbf{x}_2 := \mathbf{x}_{ini}$

**Repeat**

　　　**for** $i := 1 \rightarrow n$ **do**

　　　　　$x_{i_1} := x_{i_{\min}} + c_{gold} \cdot \left(x_{i_{\max}} - x_{i_{\min}}\right)$

　　　　　$x_{i_2} := x_{i_{\min}} + \left(1 - c_{gold}\right) \cdot \left(x_{i_{\max}} - x_{i_{\min}}\right)$

　　　　　evaluate $F_{obj}\left(\mathbf{x}_1\right), F_{obj}\left(\mathbf{x}_2\right)$

　　　　　**if** $F_{obj}\left(\mathbf{x}_1\right) \geq F_{obj}\left(\mathbf{x}_2\right)$

　　　　　　　$x_{i_{\min}} := x_{i_1}$

　　　　**Else**

　　　　　　　$x_{i_{\max}} := x_{i_2}$

　　　**end for**

**until** $\left|\mathbf{x}_{\max} - \mathbf{x}_{\min}\right| < \varepsilon$

$\mathbf{x}_{opt} := \mathbf{x}_1 \approx \mathbf{x}_2$

---

Other popular coordinate methods work with polynomial interpolation. The easiest one is Brent's method of quadratic interpolation [26]. Given three points along the coordinate direction for which

$$x_{i_1} < x_{i_2} < x_{i_3}, i \in [1,n]$$
$$F_{obj}\left(\mathbf{x}_1\right) > F_{obj}\left(\mathbf{x}_2\right) < F_{obj}\left(\mathbf{x}_3\right) \tag{8.48}$$

holds, there is a second order polynomial of the form

$$f = a \cdot x_i^2 + b \cdot x_i + c \tag{8.49}$$

that passes through the couples $\{x_{i_j}, F_{obj_j}(\mathbf{x}_j)\}, j \in [1,3]$. The minimum of the interpolated polynomial is easily determined and by this three new couples that fulfill eq. (8.48) can be selected for a better approximation of the minimum.

In principle, there is no reason that limits these approaches to the coordinate directions. In fact, these methods also work on each set of $n$ search directions that form a basis of the design space. Nevertheless, these methods are mainly used for one-dimensional optimization problems.

The simplex method by Nelder and Mead [27] is a multi-dimensional algorithm that does not use derivatives. The main idea is to build up a regular body with $n+1$ points, the simplex, in the design parameter space. For two design parameters, this will be by a triangle, for three parameters, a tetrahedron, and for even larger dimensions the equivalent hyper-body. At each point, the function is evaluated. The worst point is mirrored with respect to the geometric mean of the other points. The algorithm performs stretching and shrinking of the simplex according to Figure 8.22 while moving through the domain. In the end, the simplex collapses into one single point.

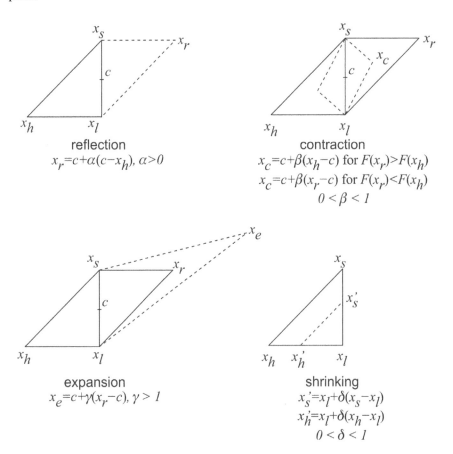

reflection
$x_r = c + \alpha(c - x_h), \ \alpha > 0$

contraction
$x_c = c + \beta(x_h - c) \ \text{for} \ F(x_r) > F(x_h)$
$x_c = c + \beta(x_r - c) \ \text{for} \ F(x_r) < F(x_h)$
$0 < \beta < 1$

expansion
$x_e = c + \gamma(x_r - c), \ \gamma > 1$

shrinking
$x_s' = x_l + \delta(x_s - x_l)$
$x_h' = x_l + \delta(x_h - x_l)$
$0 < \delta < 1$

**FIGURE 8.22** Stretching and shrinking of a simplex in two dimensions. (From Wild [22], figure 1.)

**Algorithm 8.6:** *Simplex method*

---

build initial simplex $\{\mathbf{x}_1, \mathbf{x}_2, \ldots, \mathbf{x}_{n+1}\}$

evaluate $F_{obj}(\mathbf{x}_1), F_{obj}(\mathbf{x}_2), \ldots, F_{obj}(\mathbf{x}_{n+1})$

**Repeat**

worst point: $\qquad\qquad \mathbf{x}_h := \mathbf{x}_i \mid F_{obj}(\mathbf{x}_i) > F_{obj}(\mathbf{x}_j) \forall j \neq i$

best point: $\qquad\qquad \mathbf{x}_l := \mathbf{x}_i \mid F_{obj}(\mathbf{x}_i) < F_{obj}(\mathbf{x}_j) \forall j \neq i$

second best point: $\qquad \mathbf{x}_s := \mathbf{x}_i \mid F_{obj}(\mathbf{x}_i) < F_{obj}(\mathbf{x}_j) \forall j \neq \{i, l\}$

center point: $\qquad\qquad \mathbf{x}_c = \dfrac{1}{n} \displaystyle\sum_{\substack{i=1,n+1 \\ \mathbf{x}_i \neq \mathbf{x}_h}} \mathbf{x}_i$

reflection: $\qquad\qquad \mathbf{x}_r := \mathbf{x}_c - \alpha(\mathbf{x}_h - \mathbf{x}_c), \alpha > 0$

**if** $F_{obj}(\mathbf{x}_r) \geq F_{obj}(\mathbf{x}_h)$ shrinkage: $\qquad \mathbf{x}_i := \mathbf{x}_l + \delta(\mathbf{x}_i - \mathbf{x}_l), 0 < \delta < 1$

**else if** $F_{obj}(\mathbf{x}_r) \leq F_{obj}(\mathbf{x}_s)$ expansion: $\qquad \mathbf{x}_e := \mathbf{x}_c + \gamma(\mathbf{x}_r - \mathbf{x}_c), \gamma > 1$

**else** contraction:

$\qquad$ **if** $F_{obj}(\mathbf{x}_r) \leq F_{obj}(\mathbf{x}_h)$: $\quad \mathbf{x}_{new} := \mathbf{x}_c + \beta(\mathbf{x}_r - \mathbf{x}_c), 0 < \beta < 1$

$\qquad$ **else**: $\qquad\qquad\qquad \mathbf{x}_{new} := \mathbf{x}_c + \beta(\mathbf{x}_h - \mathbf{x}_c), 0 < \beta < 1$

replace $\mathbf{x}_h$

**until** $|\mathbf{x}_h - \mathbf{x}_l| < \varepsilon$

$\mathbf{x}_{opt} := \mathbf{x}_l$

---

### 8.5.1.4 Random Methods

There is quite a large range of optimization algorithms that uses random numbers in one or the other way. The easiest implementation is the Monte Carlo search where the design variables are randomly disturbed and the objective function evaluated, and the result is only stored if the objective function is improved. From a statistical point of view, there is a good chance to eventually randomly detect the global optimum. The big shortcoming is that there is neither an assumption on the needed number of evaluations nor on the criterion to stop the procedure.

To overcome this, many random-based optimization algorithms try to model physical, chemical, or biological mechanisms that have some kind of uncertainty incorporated. At least nowadays, there are as many stochastic algorithms available as there are stochastic processes in nature. In the following, only two types of algorithms are explained, reflecting the most commonly used strategies.

The simulated annealing method [28] simulates the behavior of a liquid or metallic melt and its organization in its crystalline structures when cooling down. When the cooling process is not too fast, nature is able to find the crystalline structure, where the energy state is minimal. The interesting thing is that a thermal system has its energy $E$ probabilistically distributed among all different energy states with the so-called Boltzmann probability distribution

$$PD(E) \sim e^{-\frac{E}{kT}}, \qquad\qquad (8.50)$$

where $T$ is the actual temperature and $k$ the Boltzmann constant. The probability of low energy state increases as the temperature decreases, but also at every temperature, there is a chance to be at a state with higher energy. At the same time, the mobility of molecules decreases as the temperature of the material decreases. The probability for the system to change the energy state from $E_1$ to $E_2$ is derived from eq. (8.50) as

$$p = e^{-\frac{E_2 - E_1}{kT}}. \tag{8.51}$$

The transfer of the annealing model to the optimization problem defines the objective function value as the energy and the mobility of the molecules as the disturbance of the design variables. The most interesting thing is that due to the probability of accepting higher energy values, the algorithm is able to go uphill, thus increasing the ability to leave a local minimum.

The implementation of the algorithm is now straightforward. At each temperature, a complete Markov chain of seeding vectors is calculated and analyzed, which is a series of $n$ vectors that are disturbed randomly within one vector element. The amount of maximum disturbance is scaled with the actual temperature in order to model decreasing mobility. For each vector, the objective is evaluated and the probability of changing from one function value to the other is compared to some additional random number. After the Markov chain, the temperature is reduced by an arbitrary factor $\varepsilon_T$. With ongoing iterations, the temperature decreases down to a minimum temperature, leading to reduced perturbations and lower probability of accepting a set of design variables with higher objective function value.

**Algorithm 8.7: *Simulated annealing method***

---

$\mathbf{x} := \mathbf{x}_{ini}$
$T := T_{ini}$
evaluate $F_{obj}(\mathbf{x})$
**Repeat**
    **for** $i := 1 \rightarrow n$ :
        $\mathbf{x}_{new} := \mathbf{x}$
        $x_{new,i} := x_i + (0.5 + \text{rand}(\ )) \cdot T$
        evaluate $F_{obj}(\mathbf{x}_{new})$
        **if** $\text{rand}(\ ) < \left( p = e^{-\frac{F_{obj}(\mathbf{x}_{new}) - F_{obj}(\mathbf{x})}{kT}} \right)$ :
            $\mathbf{x} := \mathbf{x}_{new}$
            $F_{obj}(\mathbf{x}) := F_{obj}(\mathbf{x}_{new})$
    **end for**
    $T := \varepsilon_T \cdot T$
**until** $T < T_{min}$
$\mathbf{x}_{opt} := \mathbf{x}$

---

Evolutionary algorithms model the biological process of "the survival of the fittest" [29, 30], where fitness is simply replaced with the objective function value. The algorithms model the development of a population over a number of generations, where the individuals are vectors of design variables. The population now evolves based on three major principles: (a) mutation, which is the random disturbance of a component of the design vector; (b) cross-over or recombination, which is the randomly chosen mixture of two design vectors to build a new one; (c) selection by throwing away the individuals with the worst fitness.

The example algorithm shows a very simple example of such an algorithm where only one new individual is generated for each new generation. This is, of course, not the best approach but easier to understand. There are a large number of different algorithms that mainly differ in the balance between cross-over and mutation as well as the ratio between the number of new individuals and the population size. Also, within the cross-over and mutation, some algorithms work on the floating-point number level for the value of the design variables and others on the binary representation directly.

**Algorithm 8.8:** *Evolutionary algorithm*

---

Build initial population with $m$ individuals $\{\mathbf{x}_1, \mathbf{x}_2, \ldots, \mathbf{x}_m\}$
evaluate $F_{obj}(\mathbf{x}) \forall \mathbf{x} = \mathbf{x}_i, i \in [1,m]$
sort population with ascending objective function
**for** $k$ generations:
    **for** $p$ individuals:
        cross-over :
            $p_1 := \mathrm{irand}(1,m);\ p_2 := \mathrm{irand}(1,m)$
            **for** $i := 1 \rightarrow n$:
                **if** $rand(\ ) < 0.5$:       $x_{new,i} := x_{p_1,i}$
                **else:**              $x_{new,i} := x_{p_2,i}$
        mutation:
            $p := \mathrm{irand}(1,n)$
            $x_{new,p} := x_{min,p} + rand(\ ) \cdot (x_{max,p} - x_{min,p})$
        evaluate $F_{obj}(\mathbf{x}_{new})$
        **if** $F_{obj}(\mathbf{x}_{new}) < F_{obj}(\mathbf{x}_m)$: $\mathbf{x}_m := \mathbf{x}_{new}$
        resort population
    **end for**
**end for**
$\mathbf{x}_{opt} := \mathbf{x}_1$

---

Within the class of evolutionary algorithms, there has been a blurring in the distinction of a genetic algorithm and an evolutionary strategy, since both work in familiar fashion. Some researchers distinguish genetic algorithms as working on the binary representation, while evolutionary strategies use floating-point numbers.

Others define the purely genetic algorithm through the exclusive generation of individuals by cross-over, while purely evolutionary strategies employ only mutation.

## 8.5.2 DEFINITION OF THE DESIGN PROBLEM

The – most times unsolved – question is still open, namely how to define the objective function and the degrees of freedom to make the design problem well-posed and that, at the end, the achieved optimal result meets the expectations of the designer. Especially the last is crucial, as the result of the optimization is strongly depending on the formulation of the question – the composition of the objective function.

As a simple example: the general aim of aerodynamic design can be, in simple words, formulated as reducing drag while increasing lift. Giving this task to a mathematician dealing with numerical optimization, the objective function would read

$$F_{obj}(x) = C_D - k \cdot C_L \tag{8.52}$$

with a choosable weighting factor $k$. The aerodynamicist, of course, should use the aerodynamic efficiency and formulate the objective function as

$$F_{obj}(\mathbf{x}) = -\frac{C_L}{C_D}. \tag{8.53}$$

Still, both functions are valid as they achieve the required improvement, but they will result in different optima. To prove this, have a look at the minimality condition in eq. (8.31). The full gradient is given as

$$\nabla F_{obj}(\mathbf{x}) = \frac{\partial F_{obj}}{\partial C_D} \nabla C_D(\mathbf{x}) + \frac{\partial F_{obj}}{\partial C_L} \nabla C_L(\mathbf{x}). \tag{8.54}$$

The partial derivatives for the objective function of eq. (8.52) are simply

$$\begin{aligned} \frac{\partial F_{obj}}{\partial C_D} &= 1 \\ \frac{\partial F_{obj}}{\partial C_L} &= -k, \end{aligned} \tag{8.55}$$

while for the function of eq. (8.53), it reads

$$\begin{aligned} \frac{\partial F_{obj}}{\partial C_D} &= \frac{C_L}{C_D^2} \\ \frac{\partial F_{obj}}{\partial C_L} &= -\frac{1}{C_D}. \end{aligned} \tag{8.56}$$

The comparison of eq. (8.55) and eq. (8.56) reveals that the same optimum is only obtained if the weighting factor is chosen as

$$k = \left( \frac{C_D}{C_L} \right)_{min} \qquad (8.57)$$

requiring the pre-knowledge of the a priori unknown optimum solution. Therefore, achieving the aerodynamically optimum solution with the, nevertheless, mathematically correct function in the form of (8.52) is only by luck.

It is evident that using aerodynamically meaningful objective functions provides optimum solutions that fit the design problem. Using the functionals identified in Chapter 4 guarantees appropriate designs. It is important to understand the impact of different functionals on the sensitivity of the design space. Figure 8.23 shows the polars of an aircraft in different high-lift settings for landing and take-off together with the sensitivity of three different functionals associated with aerodynamic efficiency as iso-lines. Along the iso-lines, the function has a constant value. The iso-lines of the glide ratio are straight lines that intersect with the polars. With deflected high-lift system, this reveals that the glide ratio is getting largely reduced when decreasing the angle of attack. In consequence, an optimization allowing the change of angle of attack will drive into areas with low lift coefficients.

The iso-lines of the climb index are more aligned with the aircraft polars in the middle range of lift coefficients and mainly in the design range, where drag reduction is beneficial. Here, a variation of angle of attack does not lead to significant changes of the objective function, and the optimization is driven to reduce the drag in this range. An interesting functional for performance increase is also

$$F_{obj} = -\frac{C_L^2}{C_D}. \qquad (8.58)$$

It is a relation describing the relative contribution of induced drag to total drag. Especially at the very high lift coefficients, where induced drag dominates, this functional is aligned with the aircraft polar. A corresponding optimization will aim to reduce parasitic drag and/or decrease the lift-induced drag by increasing the Oswald factor of the wing, namely restoring an elliptical lift distribution with deflected high-lift system.

Still, there is another choice on how to construct the objective function. The definition of the optimization to seek for the minimum of a function calls for an adoption for functionals that should be maximized, e.g., the maximum lift coefficient or the climb gradient. Maximization can be turned into minimization in two ways, by negation

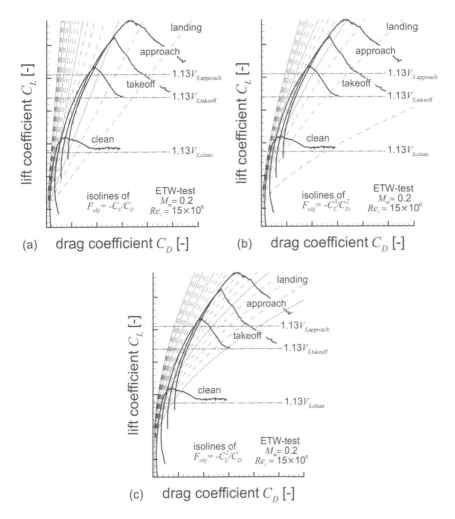

**FIGURE 8.23** Comparison of different objective functionals for aerodynamic design of the DLR-F11 high-lift configuration [3]: (a) glide ratio; (b) climb index; (c) induced drag.

$$F_{obj} = -f \qquad (8.59)$$

or, if the functional does not change sign, by inversion

$$F_{obj} = \frac{1}{f}. \qquad (8.60)$$

This additionally offers the option for functionals to be minimized to reformulate them by

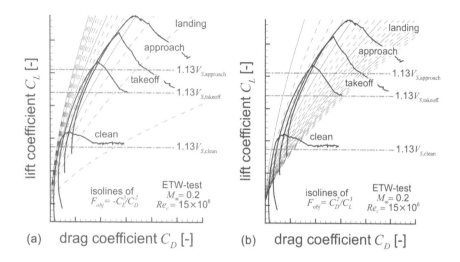

**FIGURE 8.24** Comparison of sensitivity of objective function manipulation for aerodynamic design of the DLR-F11 high-lift configuration [3]: (a) negation; (b) inversion.

$$F_{obj} = -\frac{1}{f} \qquad\qquad (8.61)$$

again, in case, when the functional does not change sign. The latter operation does not change the optimal condition but strongly changes the sensitivity of the function influencing the convergence of optimization algorithms. Figure 8.24 compares the sensitivity of the two objective functions representing the climb index $C_L^{3/2}/C_D$, which shall be maximized. Figure 8.24a uses the negation eq. (8.59), Figure 8.24b the inversion eq. (8.60). The sensitivity of the objective function is visible in the spacing between the iso-lines. The negation, in this case, shows a very low sensitivity to the right of the shown aerodynamic polars and a very strong sensitivity to the left. The inversion shows in this case a more homogenous sensitivity with only a slightly reduced sensitivity to the left of the polars. In this case, the inversion would be the preferred choice. Anyhow, this kind of sensitivity analysis should be done prior to the selection of the objective function in order to obtain an insight into the sensitivity of the design space during optimization.

Concluding the requirements from airworthiness (Chapter 3) and aircraft performance (Chapter 4), a set of functionals can be derived that directly aim to improve the aerodynamic properties without specifying additional weightings. The most important selection of functionals for construction of the objective function is listed in Table 8.1. It is truly the expertise of the designer to establish the balanced weighting of these functionals to obtain a valuable result of the numerical optimization process.

## TABLE 8.1
## Different Functionals for Improvement of High-Lift Aerodynamics

| Functional $f$ | Flow Condition | Objective(s) |
|---|---|---|
| $-C_{L,max}$ | $\alpha = \alpha_{max}$ | Minimize stall speed and derived properties, e.g. take-off speed, approach speed |
| $C_D\big/C_L$ | $C_L \leq \frac{1}{1.13^2} C_{L,max}$ | Maximize climb gradient |
| $C_D^2\big/C_L^3$ | $C_L \leq \frac{1}{1.13^2} C_{L,max}$ | Maximize climb rate |
| $C_L\big/C_D$ | $C_L \leq \frac{1}{1.23^2} C_{L,max}$ | Maximize glide path for steep descent |
| $\mu_R C_L - C_D$ | $\alpha = 0°$ | Maximize acceleration, thereby minimizing take-off run distance |

## NOTES

1. The Airbus A300 and A310 had engines mounted below the wing with a continuous slat above the engine pylon, see [6].
2. For trailing edge devices, the deflection angle is $\delta = -\delta_F$ to take into account the definition of a positive downward deflection, which is mathematically negative in x-y coordinates.
3. Wild & Brezillon [16] and Brezillon, Dwight & Wild [17] exemplarily have shown three-dimensional high-lift wing optimizations in acceptable time frames. These were obtained by highly reduced numbers of degrees of freedom, distinct spatial discretizations allowing for reduced number of computational points in the grid, or even simply coarser grids.
4. First attempts of numeric optimization of high-lift systems using RANS solvers have been made by Eyi et al. [18], Wild [19] validated the methodology. A comprehensive summary on research activities has been provided by van Dam [20].

## REFERENCES

[1] Rudolph PKC (1993) High-Lift Systems on Commercial Subsonic Airliners, NASA CR 4746.
[2] Meredith PT (1993) Viscous Phenomena Affecting High-Lift Systems and Suggestions for Future CFD Development, no. 19 in AGARD CP-515.
[3] Wild J, Brezillon J, Amoignon O, Quest J, Moens F, Quagliarella D (2011) Advanced Design by Numerical Methods and Wind-Tunnel Verification Within European High-Lift Program, Journal of Aircraft 46(1), pp. 157–167.
[4] Airbus (accessed 2019) Airport Operations – AutoCAD 3 view aircraft drawings, https://www.airbus.com/aircraft/support-services/airport-operations-and-technical-data/autocad-3-view-aircraft-drawings.html.
[5] Boeing (accessed 2019) CAD 3-View Drawings for Airport Planning Purposes, https://www.boeing.com/commercial/airports/3_view.page.
[6] Flaig A, Hilbig R (1993) High-Lift Design for Large Civil Aircraft, no. 31 in AGARD-CP-515.
[7] Woodward DS, Lean DE (1993) Where is High-Lift Today – A Review of past UK Research Programmes, no. 1 in AGARD CP 515.

[8] Tavernetti L (1992) The C-17: Modern Airlift Technology, AIAA Aerospace Design Conference 1992, AIAA Paper 92-1262.

[9] Reckzeh D (2014) Multifunctional Wing Moveables: Design of the A350XWB and the Way to Future Concepts, 29th Congress of the International Council of the Aeronautical Sciences, ICAS-2014-0133.

[10] Goldhammer M (2011) The Next Decade in Commercial Aircraft Aerodynamics – A Boeing Perspective, Aerodays 2011 Madrid, Spain

[11] Strüber H (2014) The Aerodynamic Design of the A350 XWB-900 High Lift System, 29th Congress of the International Council of the Aerospace Sciences, ICAS-2014-298.

[12] Payne FM, Wyatt GW, Bogue DR, Stoner RC (2000) High Reynolds number studies of a Boeing 777-200 high lift configuration in the NASA ARC 12-ft pressure tunnel and NASA LaRC National Transonic Facility, 18th Applied Aerodynamics Conference, AIAA paper 2000-4220.

[13] Busemann A (1935) Aerodynamische Auftrieb bei Überschallgeschwindigkeit, Luftfahrtforschung 12(6), pp. 210–220.

[14] Streit T, Wichmann G, von Knoblauch F, Campbell RL (2011) Implications of Conical Flow for Laminar Wing Design and Analysis, 29th AIAA Applied Aerodynamics Conference, AIAA paper 2011-3808.

[15] Lock RC (1964) An Equivalence Law Relating Three-and Two-Dimensional Pressure Distributions, ARC R&M 3346.

[16] Wild J, Brezillon J (2008) Optimization of 3-D Multi-Element High-Lift Device Configuration, 5th European Congress on Computational Methods in Applied Sciences and Engineering, Proceedings on CD-ROM.

[17] Brezillon J, Dwight RP, Wild J (2008) Numerical Aerodynamic Optimisation of 3D High-Lift Configurations, 26th Congress The International Council of The Aeronautical Sciences (ICAS), ICAS-2008-2.2.3, ISBN 0-9533991-9-2.

[18] Eyi S, Lee KD, Rogers SE, Kwak D (1995) High-Lift Design Optimization using the Navier-Stokes Equations, 33rd Aerospace Sciences Meeting and Exhibit, AIAA Paper 1995-0477.

[19] Wild J (2002) Validation of Numerical Optimization of High-Lift Multi-Element Airfoils based on Navier-Stokes-Equations, 20th AIAA Applied Aerodynamics Conference, AIAA Paper 2002-2939.

[20] van Dam CP (2002) The Aerodynamic Design of Multi-Element High-Lift Systems for Transport Airplanes, Progress in Aerospace Sciences 38(2), pp. 101–144.

[21] Gill PE, Murray W, Wright, MH (1993) Practical Optimization, Academic Press, London, England.

[22] Wild J (2004) Multi Objective Constrained Optimization and High Lift Device Applications, VKI Lecture Series 2004-07, Optimization Methods & Tools for Multicriteria/Multidisciplinary Design, Nov. 15-19, 2004, von Karman Institute for Fluid Dynamics, Brussels, Belgium.

[23] Shanno DF (1970). Conditioning of Quasi-Newton Methods for Function Minimization, Mathematics of Computations 24(111), pp. 647–657.

[24] Fletcher R, Reeves CM (1964) Function Minimization by Conjugate Gradients, The Computer Journal 7(2), pp. 149–154.

[25] Polak E, Ribière G (1969) Note sur la convergence de directions conjuguée, Revue française d'informatique et de recherché opérationnelle. Série rouge 3(R1), pp. 35–43.

[26] Brent RP (1973) Algorithms for Minimization without Derivatives, Prentice-Hall, Englewood Cliffs, New Jersey, pp. 195ff.

[27] Nelder JA, Mead R (1965) A Simplex Method for Function Minimization, The Computer Journal 7(4), pp. 308–313.

[28] Kirkpatrick S (1984) Optimization by Simulated Annealing: Quantitative Studies. Journal of Statistical Physics **34**(5/6), pp. 975–986.

[29] Holland JH (1975) Adaptation in Natural and Artificial Systems. University of Michigan Press, Ann Arbor, USA.

[30] Rechenberg I (1973) Evolutionsstrategie: Optimierung technischer Systeme nach Prinzipien der biologischen Evolution. Frommann-Holzberg, Stuttgart, Germany.

# Abbreviations

**A**

| | |
|---|---|
| **AEO** | All Engines Operating |
| **AFC** | Active Flow Control |
| **AFFDL** | Air Force Flight Dynamics Laboratory |
| **AG** | AGARDograph |
| **AGARD** | Advisory Group for Aerospace Research and Development |
| **AIAA** | American Institute of Aeronautics and Astronautics |
| **AIP** | Aeronautical Information Publication |
| **AMST** | Advance Medium STOL aircraft |
| **ARC** | Aeronautic Research Council |
| **ASED** | Aviation and Surface Effects Department |
| **ATC** | Air Traffic Control |
| **ATO** | Authorization to Offer |
| **AVA** | Aerodynamische Versuchsanstalt Göttingen |
| **AWIATOR** | Aircraft Wing with Advanced Technology Operation |

**B**

| | |
|---|---|
| **BLC** | Boundary Layer Control |

**C**

| | |
|---|---|
| **CAS** | Calibrated Air Speed |
| **CDA** | Continuous Descent Approach |
| **CFD** | Computational Fluid Dynamics |
| **CFR** | Code of Federal Regulations |
| **CP** | Conference Proceedings |
| **CR** | Contractor Report |
| **CS** | Certification Specification |

**D**

| | |
|---|---|
| **DES** | Detached Eddy Simulation |
| **DGLR** | Deutsche Gesellschaft für Luft- und Raumfahrt e.V. |
| **DLR** | Deutsches Zentrum für Luft- und Raunfahrt e.V. |
| **DNS** | Direct Numerical Simulation |
| **DNW** | Stiftung Deutsch-Niederländische Windkanäle |
| **DOC** | Direct Operating Costs |
| **DOW** | Dry Operating Weight |
| **DVL** | Deutsche Versuchsanstalt für Luftfahrt e.V. |

**E**

| | |
|---|---|
| **EAS** | Equivalent Air Speed |
| **EASA** | European Aviation Safety Agency |
| **EPNL** | Effective Perceived Noise Level |
| **ETW** | European Transonic Wind Tunnel |

**F**

| | |
|---|---|
| **FAA** | Federal Aviation Authority |
| **FAR** | Federal Aviation Regulations |
| **FH** | Fachhochschule |
| **FL** | Flight Level |

**H**

| | |
|---|---|
| **HiLiPW** | High-Lift Prediction Workshop |
| **HLFC** | Hybrid Laminar Flow Control |

**I**

| | |
|---|---|
| **ICAO** | International Civil Aviation Organization |
| **ICAS** | International Council of the Aeronautical Sciences |
| **ILS** | Instrument Landing System |
| **INCAS** | Institutului Naţional de Cercetare-Dezvoltare Aerospaţială "Elie Carafoli" |

**J**

| | |
|---|---|
| **JAR** | Joint Aviation Regulations |

**K**

| | |
|---|---|
| **KKK** | Kryogener Windkanal Köln |

**L**

| | |
|---|---|
| **LDLP** | Low Drag/Low Power approach |
| **LES** | Large Eddy Simulation |
| **LGE** | Landing Gear Extended |
| **LGF** | Lift Gain Factor |
| **LGR** | Landing Gear Retracted |
| **LS** | Lecture Series |
| **LTPT** | Low Turbulence Pressure Tunnel |
| **LW** | Landing Weight |

**M**

| | |
|---|---|
| **MEW** | Manufacturer's Empty Weight |
| **MLW** | Maximum Landing Weight |
| **MTOW** | Maximum Take-Off Weight |

**N**

| | |
|---|---|
| **NACA** | National Advisory Committee for Aeronautics |
| **NADC** | Naval Air Development Center |
| **NASA** | National Aeronautics and Space Administration |
| **NLR** | Nationaal Lucht- en Ruimtevaartlaboratorium |
| **NRC** | Non-Recurring Costs |
| **NTF** | National Transonic Facility |

**O**

| | |
|---|---|
| **OEI** | One Engine Inoperative |
| **OEW** | Operating Empty Weight |
| **ONERA** | Office National d'Études et de Recherches Aérospatiales |

**Q**

| | |
|---|---|
| **QSRA** | Quiet Short-Haul Research Aircraft |

**R**

| | |
|---|---|
| **R&D** | Research & Development |
| **R&M** | Report and Memoranda |
| **RAE** | Royal Aircraft Establishment |
| **RAEVAM** | Royal Aeronautical Establishment – Variable Area Mechanism |
| **RANS** | Reynolds-averaged Navier-Stokes |
| **RC** | Recurring Costs |

**S**

| | |
|---|---|
| **SAL** | Steep Approach Landing |
| **SP** | Special Publication |
| **STOL** | Short Take-Off and Landing |

**T**

| | |
|---|---|
| **TAS** | True AirSpeed |
| **TED** | Trailing Edge Device |
| **TLAR** | Top-Level Aircraft Requirement |
| **TM** | Technical Memorandum |
| **TN** | Technical Note |
| **TOD** | Take-Off Distance |
| **TOFL** | Take-Off Field Length |
| **TOW** | Take-Off Weight |
| **TR** | Technical Report |
| **TU** | Technische Universität |

**U**

| | |
|---|---|
| **UK** | United Kingdom |
| **URANS** | Unsteady Reynolds-averaged Navier-Stokes |

| | |
|---|---|
| **US** | United States |
| **USB** | Upper Surface Blowing |

**V**

| | |
|---|---|
| **VC** | Variable Camber |
| **VG** | Vortex Generator |
| **VGJ** | Vortex Generating Jet |
| **VLA** | Very Light Aircraft |

**X**

| | |
|---|---|
| **XWB** | extra Wide Body |

**Z**

| | |
|---|---|
| **ZFW** | Zero Fuel Weight |
| **ZNMF** | Zero Net Mass Flux |

# Index

**289**

For Product Safety Concerns and Information please contact our EU
representative  GPSR@taylorandfrancis.com
Taylor & Francis Verlag GmbH, Kaufingerstraße 24, 80331 München, Germany

www.ingramcontent.com/pod-product-compliance
Ingram Content Group UK Ltd.
Pitfield, Milton Keynes, MK11 3LW, UK
UKHW021117180425
457613UK00005B/125